Drone Data Metrics for Orchard Farming

Making Sense of Drone-Derived Data for Perennial Crops

AgTech Publications (Pty) Ltd

KEN TRELOAR

First Edition, 2025

Publisher: AgTech Publications

ISBN: 978-1-0492-2761-0

Cover design: Ken Treloar - AgTech Publications

Cover image courtesy: Justin Hawley, Drone Solutions International

For permissions or inquiries contact: *agtech@kentreloar.com*

Website: www.agtechdiaries.com

I dedicate this book to the farmers and the agricultural drone pilots that serve them.

I realise I'm immensely privileged to be standing at the confluence of these two important bodies. And thanks to you, I've had the opportunity to write this book.

KEN TRELOAR

Let us not become weary in doing good, for at the proper time we will reap a harvest if we do not give up.

- Galatians 6:9

KEN TRELOAR

Table of Contents

A quick note to the reader: You will see QR codes throughout this book at the end of each chapter - these will lead you straight to the corresponding blog post for further reading. Enjoy!

About the book

The ideas you will find here were first shaped during my time working at *Aerobotics*, where I have had the privilege of supporting farmers and production teams as they adopt drone data into their operations including high-stakes decision making processes.

Countless conversations in orchards and in packhouses revealed the same pattern...

Farmers want clarity.

They want to know what the technology *really* means for their orchards, groves, and vineyards - and how to separate the usual marketing promises from practical hands-on generated value.

These experiences have laid the foundation for what has come next. And in terms of my writings about drone data and agricultural technologies, the journey pretty much started on my blog, *The AgTech Diaries*.

The goal during my writing journey has been to share my enthusiasm for agricultural technology, and translate drone related concepts into clear, farmer-focused language - with education being a key component.

From there, some of the relevant posts evolved into a free week-long email course, designed to introduce readers to the fundamentals of aerial data for orchard farming.

The feedback was supportive and fairly consistent: Participants valued the insights, but expressed the need for something to accompany the coursework. Something they could keep, refer back to, and even share with their teams.

This led to the creation of the original *'Drone Data Metrics for Orchard Farming'* eBook. Built as a quick-reference companion to the course, free to access and easily shared, it was designed to get straight to the point while being fun to browse with platform screenshots and drone related imagery to help cut through the boredom that can come from lengthy documents.

While the free eBook was well received, it remains limited in what it covers. It scratches the surface, but leaves room for expansion - things unsaid.

This expanded edition of the original document that you hold in your hands today, is my answer to fill a few of the gaps and expand where I think expansion is warranted - particularly the section that covers AI-backed precision sampling, which you'll find in Chapter 9.

As such, in this humble book you will find the same foundations that lead to core metric principles, but with some added depth and detail.

Each chapter has been thoughtfully unfurled, with additional information and practical context woven into each page. At the same time, I have kept the text accessible and the insights fairly actionable.

The result is not only a companion to an email course, or a set of articles expanded and organised, but a sound reference guide with some practical utility. A book built for any perennial crop farmer, agronomist, or production decision-maker serious about using drone data with purpose and impact.

In essence, my mission of supporting farmers in viewing their orchards from above via drone data information, and to act with confidence on the insights they can gain through this process, is well underway.

The beginnings of an adventure

My first release reflects a journey that began at *Aerobotics*, evolved into time spent in the field, and which turned into the adventure that is my content creation mission. A growing set of blogs and courses, all in the name of expanding the volume of available content for industry stakeholders.

As you can probably make out, the topic of education is something I feel strongly about. Something currently lacking in the world of drone-derived insights for perennial tree and vine crop production.

I believe that knowledge grows when it's expanded, shared, and then applied. And so, this book is part of that growth trajectory.

I hope you find value in what lies within its pages.

Sincerely,
Ken

Preface

Seeing the Orchard from Above

There's a quiet revolution taking place in agriculture. And it's not in the boardroom, or in those Monday meetings. It's unfolding in orchards, where data meets dirt, and where the real decisions are being made.

This book is for those who want to farm smart.

For the ambitious and the brave.
The intelligent, the frugal, the ambitious.

It's for those who know that more information equals greater clarity. Especially when captured, processed, and applied with purpose.

From maps to meaningful action.
That's the gap we're going to be closing.

From Pixels to Progress

Purposeful action comes with knowledge and understanding. And instead of focusing on all the *inaction* out there (as this becomes wearisome) I've focused on *education*.

The mission is to address those voids you may not even realise exist. Shortening the distance between goals and achievements. In essence, to *empower*.

Perennial crops do not afford many second chances - especially within the same season. A missed irrigation issue. A hidden pest outbreak. A stressed zone left unaddressed until yield had been compromised. You get the picture.

For years, we've relied on experience, routine scouting, and gut instinct. But nowadays, there's a new tool added to the toolbox: Drone data metrics.

What's all the UAV fuss about?

Unmanned Aerial Vehicles (UAVs) and drone-based information has been around for a while now - so how are things different from a few years back?

In simple terms: The technology has advanced in leaps and bounds. And the value is ripe for the picking. Not only that, but drone-derived data has become readily available and within reach for producers big and small.

Cloud-based platforms have matured. User experience has been enhanced. And continual feedback loops have created a way for improved utility. Further to that, the steady introduction of Artificial Intelligence (AI) helps with the processing of data *and* the way we digest drone scan results. All while AI-back "smart sampling" supports the system as a catalyst for action.

While aerial insights are powerful, they're often misunderstood or under-utilised. Or worse, dismissed outright!

That's where this book comes in. It breaks down the metrics and makes them accessible. It decodes the jargon. It shows us what to measure, when it matters, and how to take the next few steps, and beyond.

More Than Just Another "Drone Book"

As mentioned, this material contains a lot that's found in The AgTech Diaries' Drone Data Metrics email course. A free mini-course designed to demystify key metrics used for orchard monitoring by drone scan.

As I still believe it's the best per-plant analysis platform out there, backed by robust machine learning processes, the *Aerobotics* software features as the main way in which I recommend you digest your drone data.

However, even if you find yourself utilising a different platform, several of the metrics (if not all) should be available in some shape or form.

And so, the content of this book remains relevant and will assist you in making best use of your drone scan results.

If you've taken my free email course already, this book deepens those lessons and provides a reference you can return to across seasons.

If you haven't taken the course, this book is your entry point. Welcome.

What this book is *not*, is simply content for content sake. It's continuity. A way to build-up knowledge, layer-by-layer, step-by-step - placed squarely onto a growing use of unmanned aerial tools in the agricultural context.

Modern, impactful, mindful, ROI-based farming.

A side-note about hardware:

As UAV hardware quickly becomes obsolete, I won't be touching on specific drone or sensor models, battery charging setups, or the like.

I've read a host of agri-drone books that while perhaps ahead of their time in terms of market adoption, and useful when published, they are dated in no time at all, by mentioning very specific drone models, and including the cost of hardware, where to buy it, etc.

The content you will find in *Drone Data Metrics for Orchard Farming* is focused solidly on the *metrics and insights* themselves.

In this way, I have aimed to build a measure of timelessness into the pages - making this a piece of value for readers in the years to come.

As long as drone-derived metrics remain relevant in agriculture, I can expand on this first edition with updated publications organically.

Keeping this in mind, remember to do your own research into procuring suitable and applicable hardware, finding suitable service providers, or about obtaining a drone pilot's licence, and how to go about it.

Alternatively, reach out to me directly over email, LinkedIn, or via the blog, if you have any questions or need a little direction on your hardware requirements.

Reading with Purpose

Whether you're a production manager, agronomist, or farm owner, the goal here is simple: to help you "read" your orchards from above; empowering you to act decisively on the data.

I know I'm at risk of sounding like a broken record here... but the simple yet underestimated act of actually *taking action* cannot be understated.

You don't need to be a tech expert - but you *do* need to care about making better decisions.

Decisions that are data-backed and supported by digital tools together with your own set of expertise and experiences.

It's a powerful combination.

Welcome to the world of Drone Data Metrics.

Enjoy the insights, but more importantly,
do something with them.

Take action.

ACTION BIAS IN AGRICULTURE: WHY FORWARD-THINKING FARMERS EMBRACE TECHNOLOGY FIRST

Chapter 1:
The Drone Data Advantage

Drone data is not reserved for tech-heads or agri-scientists. It's here, it's *available*. Moreover, it's powerful when you know how to wield it.

As renowned author on drone-related topics Louise Jupp so clearly states in her book 'Precision Farming from Above', *"The true value of drone technology lies not in the flight itself, but in the decisions that can be made from the data gathered."*.

Used with purpose, insights completely change how we view our orchards. From above *and* from within. A map of stress and health. A clear perspective on variability, and a map of production potential.

It's an orchard status summary that traditional methods simply cannot offer.

Out with the old and in with the new.
It's a matter of survival.

Target sampling ...in areas that matter most

Instead of sending scouts into the field at random, drone maps highlight the zones with the most variability. This ensures that leaf, soil, or yield samples are taken where they will reveal the greatest differences. The result is fewer wasted samples and more accurate insights about the true status of the orchard.

Prioritise scouting ...where issues are likely

Every block has its weak spots. Drone imagery brings these into focus, showing exactly where stress patterns or canopy gaps are forming. With this knowledge, we can send teams directly to the rows that need attention. This saves time, reduces labour inefficiency, and ensures problems are caught before they spread.

Segment orchards ...into meaningful zones

Not all trees are equal. Some thrive, others struggle, and many sit somewhere in between. By grouping trees into zones of similar health or vigour (or by other characteristics like canopy size) we can manage each segment appropriately.

This makes orchard variance manageable, and opens the door to practices like Variable Rate Applications (VRA) and targeted pruning programs.

Optimise input costs ...while reducing waste

Inputs are among the biggest expenses in orchard farming. Drone data helps us put every litre of water, every kilogram of fertiliser, every spray application, and labour to work where it counts. By applying resources only where they are needed, we reduce waste, protect the environment, and keep costs under tighter control. All of this adds up to an enhanced return on investment.

Plan for the future ...based on real-world data

Perhaps the most powerful advantage is the ability to build a record over time. With repeated drone surveys, we can see how decisions today affect performance tomorrow.

This cumulative dataset becomes a decision-making tool in its own right, guiding replanting strategies, capital investment, and long-term planning with a level of confidence that guesswork cannot provide.

Whether farming macadamias in South Africa, citrus in Spain, avocados in Chile, or almonds in California - the pattern remains the same: Better data leads to improved decisions. Better decisions lead to favourable outcomes.

From Maps to Management: Seeing the Unseen

Plant stress, poor drainage, pest pressure zones, and yield gaps.

These aren't always visible on the ground, and quantifying them is even harder. Especially so with tired eyes after a long day of farming. But from 100-120 metres above, these trends reveal themselves - and with a surprising level of precision.

Additionally, drones make the invisible, visible. Patterns that are difficult to near impossible to see with the naked eye. As such, the data drones collect highlight patterns that help us take targeted, meaningful action.

From macro patterns to granular insights, the view from above tells the story. And to be clear - using drone data is not about replacing expertise. It's about *enhancing it* with layers of high-value clarity.

Quick Case Study:
Jaco's turning point

A seasoned macadamia farmer in South Africa, Jaco used to rely entirely on visual scouting. He had a strong instinct for reading his trees. But he also had a feeling he was missing early signs of trouble. Turns out, a faulty irrigation valve had gone unnoticed. Affecting over a third of the block. Starving trees of water and nutrients.

The data from the survey picked this up. Checks on the ground confirmed it. He fixed the issue. And yields bounced back to what he expected. "Without the detailed map," he said, "I would only have noticed when it was much too late." Data didn't replace his experience. It amplified it.

In other words, drone data helps us spend less, act faster, and farm smarter. Catching issues before they become worse. Helping us to locate and define limiting factors; while supporting remedial actions for enhanced production and optimal yields.

Drone Data Info Recap

This chapter was our humble launchpad into the world of precision insights.

Next, we'll explore what I believe are some of the most important metrics available today, how to interpret them, and how to move from image to action.

1. Drone data gives us a bird's-eye view of variability across the orchard. And faster than any scout could ever do on foot.
2. A single flight can reveal stress patterns, canopy trends, and risk zones invisible to the human eye.
3. It's not about collecting everything; it's about capturing the right layers to guide smarter decisions.
4. Good data shows us where to look, what to prioritise, and what to do next.

5. Timing matters. Survey before symptoms show, not after the damage is done; planning your drone scans around crop phenology to get the most ROI.

6. Think of drone maps as conversation starters. Not always exact answers, but prompts for smarter and enhanced investigation.

7. Drone data isn't the be-all and end-all solution. It's the lens that sharpens our thinking. It's an action enhancer.

Each drone derived image is a snapshot of potential, and only as powerful as our ability to understand it.

A map on its own is only an image, but when we know what went into capturing it - the drone type, the sensor payload, and background on the planning behind each flight - we start to understand and appreciate it that much more.

In the pages to follow we'll cover each of the key data metrics types that support advanced orchard analysis. You're probably chomping at the bit to get started - and you'd be right to be excited.

But first, let's cover some of the foundations: exploring the basics of the hardware, sensors, and strategies that make all the UAV insights possible.

DIGITAL FIELDS, SMARTER YIELDS: HOW BIG DATA IS REDEFINING MODERN FARMING

Chapter 2:
Key Data Metrics & UAVs

Drones, Sensors, and Strategy Explained

UAVs (Unmanned Aerial Vehicles) are transforming agriculture. Once considered cutting-edge, they are now reachable tools in the precision agriculture toolkit.

The ability to collect high-resolution data, rapidly and repeatedly, means that drones are now indispensable. Even more so while navigating the complex nature of decision-making and task-execution required for effective perennial crop production.

That being said, new data sources are only truly useful if they help us to make better farming decisions. In short, sound decision making is based on a combination of available knowledge *and* critical thinking.

Knowledge provides the raw information being considered, while critical thinking is the process used to analyse, evaluate, and synthesise the influx of information to reach a well-reasoned conclusion. This process helps us to make informed choices by objectively assessing evidence, questioning assumptions, and weighing potential outcomes rather than relying solely on emotion or intuition.

Hence, there is value in understanding the basics of the machines and sensors that make aerial data possible. Without this knowledge, it's easy to fumble drone survey timing, misinterpret results, under-value or miss important insights, or fall prey to a service that promises more [or less - believe it or not] than it can deliver.

Knowing the difference between drone types, sensor options, and flight strategies helps us to ask sharper questions when liaising with service providers or working with colleagues - while providing added confidence when comparing quotes or project proposals.

Additionally, we end up armed with clarity; able to manage expectations between ourselves, agronomists, consultants, or technology providers.

Knowledge keeps the wool from being pulled over our eyes. And it's the farmer's leverage in an industry where technological ideologies often run ahead of practical outcomes.

At the end of the day, producers who understand these fundamentals make better choices. They can judge whether a multirotor drone will capture the detail they need in a high-value orchard, or whether a fixed-wing platform is best suited for wide-area monitoring.

They can recognise when a simple RGB scan is enough, or when it's worth investing in multispectral, thermal imagery, or all the full suite of metrics. Most importantly, they can link each data type and its utility back to farming thought processes around pruning, irrigation, replanting, and input planning - among others.

It's about perspective. We don't need to become drone pilots ourselves, but by grasping the essentials we ensure drone data remains a powerful tool: serving our orchards, our teams, and our goals.

With the right knowledge, we stay in control.

Remember, drones not only take "pretty pictures", but reveal problems before they become costly mistakes and misfortunes. This chapter lays the foundation for understanding how drones support digital farming.

Layers That Build Leverage

As we've covered, the focus of this book is placed firmly on drones for surveying; concentrating on drones used for mapping, monitoring, and data collection over orchards and vineyards.

As you can imagine, these drone missions differ from spraying, delivery, or any other kinds of drone-related practice.

The emphasis here is on the production of perennial crops - like fruit and nut farming - where added precision holds value for multi-year growing assets that will be "stuck in the ground" for decades at the least.

We shine a spotlight on both near-term results and long-term performance trends. Information that not only informs strategic decisions, but which fuels your operations in a sustainable way.

In a nutshell, it's about business and the core assets are *alive*. This makes farming a unique and complex industry where biological life forms are the primary resource.

Unlike most other businesses, this means farming requires managing *living things* - like vines and trees - while managing intrinsically *unpredictable variables* such as weather, pests and disease.

A Story of Fundamentals and Strategy

The ***fundamental* advantage** of agricultural drones used for mapping lies in their ability to provide near real-time information used to navigate farming business challenges - right from the very first survey.

Whether it's adjusting irrigation scheduling, applying targeted fertilisers, or identifying early signs of pest pressure, mapping drones allow us to act earlier and with added precision.

After the initial surveys, the ***strategic* advantage** of agricultural drones comes from building a cumulative record over time. Each scan adds another layer. Each season adds another chapter. As trends emerge... patterns of growth, stress, and variance reveal themselves in ways that single surveys cannot provide. Employing a good balance of these two strategies supports stronger planning, and increased accuracy on forecasting and monitoring - an improved alignment of orchard practices with strategic goals.

Basic Drone Types: Going from Flight to Insight

These fundamental and strategic advantages emerge from the data metrics derived from onboard sensors mapping drones carry.

Sensors (the mapping drone's most important payload) collects data in several spectral bands.

This data, once processed, becomes a series of practical insights. Guiding operational decisions with a level of clarity once out of reach for most producers.

The agricultural drones that carry these sensors generally come in three main formats. Each setup suits a certain set of conditions like: where the survey is taking place (terrain), the environmental conditions (weather), hardware capabilities, the scope of work (flight time / acreage), operational goals (turnaround / timing), desired outcomes, and budget.

1. Fixed-Wing Drones

Fixed-wing drones excel in endurance and efficiency, making them ideal for large-scale orchard surveys.

By relying on aerodynamic lift rather than rotors, fixed-wing UAvs can remain aloft for extended periods and cover vast areas in a single flight (up to hundreds of hectares) without frequent battery changes. This makes them the go-to platform when coverage is king.

Equally important is their ability to carry heavier payloads. Whether it's multispectral cameras, LiDAR units, or additional sensors, fixed-wing platforms handle greater weight more stably, enabling more advanced data capture per mission.

That said, fixed-wing drones demand open launch and landing zones, and cannot hover. You trade manoeuvrability for efficiency. Planning flights takes technical know-how, and operations in rugged or closed-canopy areas are challenging.

2. Multirotor Drones

Multirotor drones (quadcopters, hexacopters, and octocopters) stand out for their versatility and precision. Their ability to hover, take off quickly, and navigate smoothly through complex orchard terrain makes them highly effective for surveying and mapping.

This agility allows them to capture detailed data at lower altitudes, or makes them handy for quick single block surveys; lending themselves well to fast and flexible on-the-go scans as required.

Additionally, multirotor drones are particularly valuable when operating in orchards with irregular shapes or mixed terrain. Because multirotors can fly slowly and hover in place, they are able to capture overlapping imagery with high accuracy. This results in high-resolution orthomosaics and per-tree data that are essential for orchard management.

Modern multirotors often come equipped with stabilised gimbals and RTK positioning, which ensures consistent image capture and precise geo-referencing.

The trade-off comes in endurance. Multirotor drones generally have shorter flight times and smaller payload capacities compared to fixed-wing platforms. This limits their efficiency for very large survey areas, where multiple flights may be required.

However, for small to medium-sized orchards, or for detailed surveys within larger estates, in varying terrain - multirotors remain the most practical and flexible choice.

3. Vertical Take-off & Landing (VTOLs)

These drones combine the efficiency of fixed-wing platforms with the flexibility of multirotors. VTOLs can take off vertically, which removes the need for a runway, and then transition into forward flight for greater endurance and coverage. This design allows them to handle large survey areas while still offering the ability to hover for targeted data capture.

Their dual capability makes them especially useful in mixed farming environments. They can map thousands of hectares in a single mission, yet still manoeuvre in and out of tighter orchard spaces when precision is needed.

This makes VTOL drones a strong option for producers with both expansive blocks and more complex or fragmented portions of land.

The trade-off is complexity. VTOL systems are mechanically more advanced and often come at a higher price compared to either fixed-wing or multirotor drones.

They require greater operator skill, more maintenance, and careful planning to get the best out of each mission.

Despite this, they remain a powerful tool when flexibility across varying conditions is a priority.

Quick Summary:

Fixed-wing: Exceptional range and payload capacity, efficient for large-scale surveys, but requires open space for take-off and landing and cannot hover.

Multirotor: Highly precise and agile, ideal for small to medium orchards or detailed surveys, but limited in endurance and area coverage.

VTOL: A strong hybrid option, combining the range of fixed-wing with the flexibility of multirotors, but more complex, costly, and demanding to operate.

Hardware, Sensors, and Pre-Flight Strategy

The right drone for a job depends on many factors: terrain, target crop, urgency, weather, and the level of detail required. Just like any farm tool, it's about choosing what works best for the task at hand. Hardware and sensors included.

Hardware: The UAV Platform

The drone is the foundation of any survey mission. Endurance, payload capacity, manoeuvrability, and reliability all matter.

A fixed-wing drone may be the better choice for large orchards that demand long flight times, while a multirotor is more suited to smaller or irregular orchards that require agility. VTOL drones sit in between, offering versatility but at higher cost and complexity.

Battery systems, frame durability, and flight control electronics all influence how consistently the drone can deliver quality survey data. All of these factors must be taken into account, and any pilot worth their salt will be doing this already.

Sensors: Payloads Capturing Data

Sensor choice is central to the data collection strategy. The sensor is the component that detects and records electromagnetic radiation, whether in the visible spectrum or beyond.

Standard RGB cameras provide high-resolution "true colour" imagery, while multispectral sensors extend into near-infrared and red-edge bands for vegetation indices like NDVI and NDRE. Thermal sensors capture heat signatures that reveal transpiration and irrigation performance.

Some UAV systems allow for modular payload configurations and sensors can be swapped in and out, but keep in mind that while this increases flexibility it adds operational complexity. Aim to get as close to an all-in-one setup solution as possible.

Outside of astute and thorough flight planning, and integrating AI for data processing, this decision around drone and sensor configuration is one of the most important, and will simplify operations and ensure a high level of survey efficiency.

Pre-flight Strategy: The Operational Layer

Even the best hardware and sensors will fail to deliver if missions are poorly planned. Pre-flight strategy is where the value of a survey is either secured or lost.

While the type of crop and the insights we want to capture shape the flight plan. Factors such as terrain and weather must always be considered; since wind, temperature, and uneven ground can affect flight stability and data quality.

Practical details matter just as much. Battery life (flight time) and memory capacity all determine how much ground can be covered in a single mission. Ease of sensor swapping can make the difference between a smooth workflow and wasted time in the field.

Finally, an element of redundancy and support is essential. Having backup equipment, extra batteries, or technical assistance close at hand, ensures that if something does go wrong, the mission does not become a lost opportunity.

These are some of the operational layers and logistics behind each drone survey that will need to be considered. Careful preparation ensures that the resulting dataset is consistent, reliable, and aligned with the goals of the operation.

Whether conducting the surveys yourself or relying on a service provider, it's still important to understand these factors. Knowledge and understanding helps us to set expectations, ask the right questions, and ensure the service received is fit for purpose.

Data Collection Approaches

Getting the job done in the way that makes the most sense.

There are several routes we can take to access drone-derived data - each with its own balance of cost, convenience, and control.

Working with **an end-to-end service company** is one option. In this model, a provider handles the entire workflow: the drone flights, the processing of imagery, and the delivery of insights through a digital platform.

It's the fully serviced package, requiring little to no technical knowledge from the farmer in terms of data collection. The advantage is convenience and speed - but note that this option usually comes at a premium.

Farmers need to trust the provider's consistency and responsiveness. As such, this premium cost of all-inclusive surveying must come with "premium" service. Keep this in mind, and if the service standards are not up to scratch, switch providers sooner rather than later.

Another approach is to **contract a local drone pilot,** and to process the data separately through a third-party analytics provider. This can be a good option where a trusted pilot is available in the area, and if it makes sense in terms of budget.

The self-serviced option has enhanced flexibility compared to pre-planned fully serviced drone scans, which are often planned weeks or months in advance. Local providers may be able to fit in a survey for you at short notice, or change the scope of the drone scan as needed with little to no fuss - as long as both parties are clear on any cost related repercussions.

While this option has added flexibility and allows us to choose the analytics platform that best suits our needs - it does mean managing two relationships (and two line items when it comes to budgeting). One with the pilot, and another with the processing company. Keep this in mind if you go this route.

Another option involves choosing to **develop in-house data capture capacity**, purchasing your own drones and training a team member or two as pilots. Data capture is done internally, while processing is outsourced to a specialist platform.

This approach suits operations that plan to survey regularly, and want more control over the timing and frequency of drone surveys. This approach helps to avoid taking on the complexity of building a data processing pipeline, as well as the need to liaise constantly with service providers who may have other work orders, clients, and allegiances.

The final option is the **100% in-house model**, where everything from flying to data processing, software development, and everything in between, is handled internally. This approach is far less common, since it requires high levels of technical skill, significant capital investment.

The in-house option is typically feasible for large-scale agribusinesses or research institutions exclusively, with resources to run a robust drone operations ecosystem, ongoing hardware support, software development, data science functions, all under one roof.

Output representations: Going from imagery to insight

Drone surveys produce a large volume of raw imagery, and much of the value lies in how that imagery is processed into useful representations of the data.

These outputs can be grouped into three broad formats, each with its own level of detail and application.

1. Visuals & Map layers
2. Zonal representations
3. Individual plant data

The map layer (orthomosaic map) is created by stitching hundreds or even thousands of overlapping drone images into one seamless, geo-referenced picture of the orchard as viewed from above. A bird's-eye perspective that is accurate in scale and location, allowing us to see the entire survey area in one view.

Orthomosaics are ideal for the standard top-down overview, identifying large-scale patterns (areas of general decline, wind damaged spots, flooding, areas of poor establishment, cover-crop prevalence, changes in terrain post weather event, so forth and so forth) and for building a visual baseline that can be compared over time.

The zonal view, where processed data is aggregated into meaningful areas or segments. Instead of looking at raw layer data, these are broken up into meaningful zones of similar performance or characteristics.

Another way data platforms go about creating these representative zones is to take the outputs for every single tree, and using per-tree information to group the like trees into zones.

This is arguably a better way of utilising the initial orthomosaic information as per-tree data cuts out the outside noise - like vegetation in interrows or between trees. This “noise” may otherwise be included in zonal outputs if general observations taken from the map layers are used exclusively to create zonal representations.

In general, utilising zones makes interpreting variability across a block quick and easy, no matter the metric, and they are particularly useful in guiding sampling strategies or when implementing variable rate strategies.

In essence, zonal layers simplify decision-making, especially where per-tree data and orchard size can be overwhelming.

The third, and often most detailed output, is **plant-level data**. Here, each individual tree, bush, or vine is analysed and displayed, with metrics such as canopy size, health stats, or thermal info linked directly to a GPS location.

This level of granularity is powerful for orchard management, as it highlights outliers (for example, the weak or exceptional performers) and allows for precise interventions. Per-plant information creates a digital inventory of every asset, and the ability to have each of them tracked season by season.

Modern platforms such as those developed by *Aerobotics* offer all three formats in a single system. Farmers can zoom out for the big picture across the farm, zoom into zones for management insights, and finally drill down to individual trees when specific action is required.

These three visualisation formats - data layers, zonal views, and per-plant metrics - can also be compared *alongside* one another for added diagnostic utility.

This ability to move seamlessly between formats and different scales of detail makes drone mapping data practical, flexible, and highly actionable for orchard farmers.

Introduction to The Core Metric Types

All metric types are derived from sensor data. In basic terms, the light or heat captured by the drone's onboard camera system.

While there are several multispectral indices used in agriculture, as well as differing thermal metrics, the primary types of outputs we will cover in this book are:

- RGB visuals
- Digital Elevation Models (DEMs)
- Multispectral information
- Canopy metrics
- Plant census statistics
- Thermal-based information
- AI-generated waypoints

RGB imagery refers to remote sensing images that reflect what the human eye naturally sees. Often called "true colour" or "natural colour" imagery, these images display landscapes in a way that looks familiar to us - making them a foundational and intuitive layer for agricultural monitoring.

Some RGB sensors may also incorporate near-infrared data to enhance vegetation detail, especially in green areas. This improves the clarity and visual accuracy of plant health assessments.

The technology behind RGB imagery is conceptually similar to how printers combine colours using models like CMYK. In digital sensors, combinations of visible light bands are processed to create realistic outputs.

Beyond visual assessment, RGB imagery also plays a technical role. Through techniques like stereo photogrammetry, RGB information produces **Digital Surface Models** (DEMs) - helping to map elevation and gradient information without the use of LiDAR.

These dual functions (visual clarity and spatial data generation) make RGB a critical building block in the agricultural drone data information stack.

Multispectral imagery captures data within specific wavelength ranges across the electromagnetic spectrum. Wavelengths are either separated by filters or by sensors set to record very particular wavelengths. These sensors detect frequencies beyond the visible range of light, such as infrared and ultraviolet - capturing information we cannot see with the naked eye.

Multispectral imaging measures light in a small number of spectral bands. Typically three to fifteen. Hyperspectral imaging on the other hand, is a special type of spectral imaging with hundreds of contiguous bands. More common in research than in farming practice - so we will only cover Multispectral data use within the scope of the chapters to follow.

A widely used multispectral metric is the Normalised Difference Vegetation Index (NDVI) which you may have heard of already. This index provides insights into vegetative vigour and vitality. Based on capturing the reflectance of near-infrared light.

NDVI is just one of the multispectral data types. Another is NDRE - a powerful index which we will explore in finer detail in Chapter 5.

Canopy area metrics, while far less glamorous, are among the most valuable. Using basic RGB imagery, data such as per-tree canopy surface area and indicative vegetative volume can be generated. These are vital statistics to have on hand when paired with metrics like tree health and stress.

Plant census statistics provide an accurate inventory of orchard assets. Replacing slow, error-prone manual counting with fast, reliable, AI-assisted analysis.

By identifying each tree, and locating gaps or decline patterns, these tools improve replanting decisions, land use efficiency, insurance claims, and budget planning.

Next, **thermal data** can be used to measure transpiration - how well a tree takes in and utilises water. Healthy trees release water at canopy level and maintain cooler temperatures. High-stress trees exhibit hotter canopies. Mapping this for each plant shows with impressive precision exactly where issues and problems exist.

Thermal metrics help guide irrigation schedule protocols, reveal faulty emitters, mark frost-risk zones, and help place soil probes where they will provide the most representative data.

Finally, **AI-generated waypoints** transform insights into actions with the help of advanced models and weighted extrapolations.

While these outputs are more a *result* of metrics rather than a pure metric itself - I feel compelled to mention AI-generated waypoints within the realm of precision sampling.

AI continues to play an increasingly bigger role when it comes to converting data outputs into tertiary information adapted and applied for impact-rich insights and tailored in-field actions.

By flagging representative or anomalous trees, these data-derived AI-backed waypoints guide teams on the ground directly to the plants of interest or those of most concern.

This technology ties a bow around the relationship between aerial scan outputs and precise in-field management strategies.

Up next...

Now that we've laid out the building blocks of drone use in agriculture - the drone types, the sensors, the strategies, and the outputs - it's time to dive into each of the main metric types.

Knowledge of the metrics is not only about building the confidence to ask the right questions, and to judge the value of what is offered, but to connect drone sensor capabilities back to real decisions we want to make, and how these play out in the field.

Now that we understand some basics around how drones capture data, and in what format the information is delivered, it's time to look at the first and most accessible of the metric types: RGB imagery.

Often dismissed as "just pictures," RGB remains the foundation of aerial insights, and now that it's such a common output, it's severely underrated. In the next few pages, we'll see why, and how, RGB images can reveal more than meets the eye.

FROM PIXELS TO PRIORITIES:
USING DRONE SURVEYS TO MANAGE YOUR ORCHARDS

Chapter 3: **RGB Imagery**

Patterns Revealed Through Light and Colour

What looks ordinary to the naked eye can become a powerful decision-making tool when viewed from above.

RGB stands for "Red, Green, Blue". The three primary colours of light that our eyes can see.

These are the colours that the three types of cone cells in the human eye are sensitive to. By mixing these three colours of light in different combinations, we can see a wide range of additional colours, including white light when all three are combined equally.

As such, by combining Red, Green, and Blue in varying intensities, it's possible to reproduce the entire spectrum of visible colours.

In digital imaging, RGB is the standard method for capturing and displaying colour, used in everything from smartphone cameras to professional-grade sensors.

RGB imagery is the most commonly available type of aerial data captured by drones. In fact, most drones - from entry-level consumer models to professional agricultural platforms - carry RGB cameras as standard.

Because RGB imagery looks "normal", it's underestimated in precision agriculture. However, the aerial perspective it provides can reveal patterns, contrasts, and irregularities that are near impossible to see from the perspective of being at ground level.

Subtle differences in canopy colour or density. Gaps in planting or missing trees. Patterns of decline across an orchard block. Waterlogged areas, erosion patterns, or the extent of wind damage.

These observations form the foundation for steps toward more advanced analysis.

Even without specialised sensors, RGB imagery delivers actionable insights - particularly when combined with AI-driven data processing.

While more specialised sensors such as multispectral and thermal provide additional insights, RGB imagery offers a versatile starting point for agricultural analysis.

How RGB sensors capture colour

An imaging sensor, whether CCD (Charge-Coupled Device) or CMOS (Complementary Metal-Oxide-Semiconductor) does not naturally detect colour. Instead, it measures the intensity of incoming light.

In most cases, to “see” colour digitally, a filter known as a *Bayer filter* is placed over the sensor.

A Bayer filter is a grid of red, green, and blue filters arranged in a repeating 2×2 pattern:

- 1 × **R**ed filter
- 2 × **G**reen filters
- 1 × **B**lue filter

Each pixel on the sensor is covered by one of these filters. There are two green filters because our eyes are more sensitive to green light.

The raw output of Bayer-filter cameras is referred to as a Bayer pattern image. Since each pixel is filtered to record only one of three colors, the data from each pixel cannot fully specify each of the red, green, and blue values on its own. Therefore, to obtain a full-color image, various algorithms can be used to interpolate a set of complete red, green, and blue values for each pixel.

These algorithms make use of the surrounding pixels of the corresponding colors to estimate the values for a particular pixel. For example, a pixel under a green filter will only record green light, and the camera's processor reconstructs a full-colour image by combining data from adjacent pixels.

This process is known as *demosaicing*. The end result of using demosaicing algorithms is a high-resolution colour image, and one that closely matches what the human eye perceives.

Why colour is only the first step

The quality of each image also depends on how steadily the sensor can record it during flight. Without stability it's hard to make sense of any recordings.

As we know, drones operate in variable conditions (wind, turbulence, and rapid manoeuvres), all of which can introduce blur or distortion. For agricultural mapping, and in any situation where precision matters, *stabilisation* is just as important as the sensor itself.

Most agricultural drones mount their cameras on a stabilising gimbal. This three-axis support system actively counters vibrations and movement in real time, keeping the camera steady and level throughout mapping.

The three-axis gimbal is an evolved form of stabilisation technology that grew from early concepts like those in the 19th century with the invention of the gyroscope.

The development of the gimbal is a refinement of this principle, evolving from the mechanical and electronic stabilisation used in naval and aerospace applications, now seen in modern consumer electronics.

Today, gimbals are widely used in photography and videography for their ability to stabilise footage by using pitch, yaw, and roll (the three axis) driven by brushless motors that counteract unwanted movements.

By isolating the sensor from drone motion, gimbals ensure that each image is sharp, consistent, and ready for stitching and further processing.

In practice this means cleaner orthomosaics, reliable per-tree analysis, and greater confidence in the outputs derived from RGB imagery obtained during mapping flights.

Sharper imagery for precise measurement and mapping, consistent perspective across images, and reduced distortion even during less-than-ideal flying conditions. The gimbal is one of the unsung heroes in the drone-tech arsenal.

The quality of the raw RGB data is critical because it forms a base layer for generating downstream outputs.

Every orthomosaic, every digital elevation model, and the accuracy of plant annotations rely in most cases on sharp, crisp, and impeccably aligned RGB imagery. If the base layer is flawed, every subsequent layer could carry that flaw.

This is why high-quality RGB capture is not only about recording a “pretty picture” from above, but about recording one that’s clear and precise.

Clear imagery is about securing a reliable foundation on which the rest of the data analysis can stand firm. And thus, an important building block within the drone data capture process.

When RGB is done well, it anchors much of the data stack, ensuring that insights around tree counts, canopy growth, and gradient are trustworthy and reliable.

Can you conduct drone scans for a non-RGB survey - skipping the RGB layer completely? Yes - but the rules of stability apply regardless.

While some multispectral or thermal camera sensors utilise a gimbal, some do not. Fortunately, the very nature of the UAV platform itself offers a certain level of stability. Another reason why drones used for mapping are such a great choice.

Compared to manned or unmanned ground based vehicles (UGVs), drones are unmatched as they are not restrained by having to track the terrain physically.

Although drones (UAVs) have wind and weather to contend with, they adjust speed and flight angles to compensate - in a way becoming the gimbal themselves - with pitch, yaw, and roll.

A pilot's mindfulness about terrain changes, weather, flight parameters, and the hardware required to get the job done correctly, all affect the stability of the flight and image capture, and ultimately the data capture results.

At this point I have to mention that it's important to work with an expert service provider. Choose a data-capture partner you can trust, who can supply you with examples of their work and reliable references. And if you're flying yourself, do the required training to make sure that any data capture (no matter the sensor) is done in a professional manner to ensure the best results.

Practical Uses for RGB Imagery

While the visual appeal of RGB data is clear, its real value lies in analysis and decision-making. Six core instances stand out:

- **A bird's-eye view.**

  An aerial perspective shows the entire orchard at once, putting every row and block into context.

  This makes it easy to spot irregular layouts, bottlenecks in access routes, or areas where natural features like rivers or slopes affect the crop.

  Seeing the orchard in its entirety also helps with communicating plans to staff, consultants, or investors.

- **See macro trends and patterns.**

  Differences in colour or canopy density across blocks can point to nutrient variation, irrigation inconsistencies, or pest pressure.

  These broad patterns may not be visible from the ground, where focus tends to be on smaller details. Identifying macro trends helps farmers ask better questions and dig deeper where needed.

- **Identify obvious problem areas.**

  RGB maps make issues stand out clearly - patches of dead or missing trees, areas hit by wind or hail, or young trees that were never established properly. These areas can then be marked for further investigation.

  Without aerial imagery, these problems often remain hidden until yield losses show up at harvest.

- **Compare with other datasets.**

  RGB imagery provides the visual baseline against which other data layers can be tested.

  For example, a yellowing patch of canopies in the RGB map may correspond with high stress as depicted in multispectral indices, or align with a slope change revealed with elevation data.

  Cross-referencing builds confidence in diagnosing issues and reduces the risk of misinterpretations taken from a single dataset.

RGB imagery also gains (and adds) more value when viewed alongside some of the other data metrics - offering opportunities for enhanced diagnosis.

These additional data metrics form part of those we'll cover in greater detail later in the book - but here's a sneak-peek:

Multispectral imagery adds a deeper layer of insight. While RGB can show colour variation, multispectral analysis confirms whether those visual differences relate to actual plant stress or nutrient imbalances.

Indices such as NDVI and NDRE detect subtle changes in chlorophyll and canopy vigour, helping distinguish between superficial colour differences and true physiological problems.

Digital Elevation Models (DEMs) provide the physical context for what RGB imagery shows. Topography influences drainage, erosion, and soil depth, which in turn affect tree health.

Overlaying RGB with a DEM can explain why certain rows lag behind others, or why stress consistently appears in low-lying areas. This context turns a flat image into a spatially informed map that reflects the reality of the land.

Thermal imagery strengthens the diagnosis further by showing how plants are using water. Even if RGB indicates uniform canopy cover, thermal scans can reveal which areas are transpiring efficiently and which are under stress.

Thermal based information makes it possible to link visible anomalies with irrigation faults, blocked emitters, or underlying soil moisture issues.

- **Track changes over time.**

  When drone surveys are repeated across seasons, RGB maps create a visual record of progress. A block that was replanted after frost can be checked for uniform growth the following year. A section where fertiliser practices were adjusted can be compared before and after. This historical record builds into a valuable management tool that shows whether interventions are working.

  Keeping in mind that RGB imagery is comparatively inexpensive to collect, it's well-suited to regular monitoring. As such, one of the most powerful uses of RGB imagery is to in fact track changes over time.

  Repeat follow-on surveys allow us to measure canopy growth following replanting, assess the impact of pruning or thinning, monitor the success of drainage or irrigation improvements, and detect emerging problem areas before they escalate.

Of course, this is a non-exhaustive list of uses, and each arming situation will afford additional or fewer opportunities for analysis in terms of RGB imagery and tracking changes over time.

- **Overlay operational data.** One of the most powerful uses of RGB imagery is as a backdrop for practical farm information. Irrigation layouts, drainage lines, soil sampling grids, or even harvest records can be layered on top. This brings operational context directly onto the orchard map, helping farmers see how management systems interact with what's happening in the field.

RGB imagery may seem basic, but its power lies in providing a reliable visual foundation. It offers both a clear overview and a flexible platform for integrating other data layers.

By turning pictures into context, and context into action, RGB scans set the stage for much deeper analysis.

While human eyes are naturally good at spotting differences, quantifying those differences by hand is slow and unreliable.

This is where AI-powered processing adds value; transforming RGB imagery from "interesting pictures" into hard, actionable data:

- Count plants automatically.
- Detect gaps in planting rows.
- Measure canopy area and volume.
- Generate Digital Elevation Models.
- Compare changes between seasons.

- Identify vegetation anomalies visually.
- Notice terrain characteristics from above.
- Correlate and contrast with additional datasets.

This shift from visual interpretation to quantified measurement - especially with the aid of machine learning processes - transforms RGB imagery into hard, actionable data you simply cannot ignore or neglect.

Quick Case Study:
Brendan's low-lying block

Brendan's orchard looked healthy at first glance in the RGB survey. The canopy appeared uniform, and there were no obvious gaps or die-off.

However, one corner displayed canopies with a lighter shade of green, a few yellowing. When Brendan layered the RGB insights with the Digital Elevation Model, it was clear that this was a low-lying corner of the block. Water was likely pooling there after heavy rains, with nowhere else to go.

The Multispectral data layer quickly confirmed this gut-feel assumption. Stress levels in that section of the orchard were consistently higher than the rest of the block. Vigour was down.

Thermal scans further backed this up, showing lower transpiration values in the same rows, compared to the rest of the block even though they are all on the same watering schedule, and canopy areas are similar.

With the evidence stacked across three additional data layers, Brendan knew what he had to do. He sampled the area with precisions - testing the soil, and gauging compaction - and actioned targeted drainage and soil works protocols as needed.

He took things a step further, and made sure to flush irrigation lines on the lower side of the orchard, and replaced faulty emitters in the affected zone, and throughout the orchard. If infiltration issues were only part of the problem, he'd make sure that water delivery was also taken care of.

Additionally, since per-tree information could tell Brendan about outlier trees outside of the primary zone of concern (that low-lying corner) he could spot and mark them digitally with ease, create a sampling route, take action, and make best use of time in this orchard.

The result? The intervention worked. By the next survey, the previously struggling corner of the orchard had begun to recover. And overall variance values across all metrics were down too - all due to targeted action in the worst-off part of the orchard, and some added work on outlier trees.

In this case, using RGB as a gateway metric to leveraging additional insights, proved that there is value in *combining datasets* for sharper diagnosis and expert action. While tracking progress over time provided the confirmation that efforts were moving tree health in the right direction.

The foundation layer

RGB imagery may look simple, but to reiterate: it underpins everything else. It's the entry point for most farmers into the world of drone derived data - the first map that makes sense without much explanation, and the layer on which many other metrics and actions can be built on and leveraged.

Without sharp, consistent RGB data, every follow-on analysis (whether multispectral, thermal, or elevation based) is built on shaky ground. Take the time to give it its due attention.

RGB is a gift of clarity at both the whole-orchard scale and at the level of individual trees. It provides context, it highlights patterns, and builds a baseline that can be measured against over time.

More importantly, it makes technology accessible. Anyone can read an RGB map, which is why it plays such a central role in farmer adoption of aerial insights.

When we treat RGB as more than just a picture, it becomes the foundation of a reliable data solution stack. It anchors additional datasets, validates trends, and gives us confidence that what we see is the real situation on the ground.

Done well, RGB turns basic aerial orchard photography into an integral metric within a powerful management multitool. Do not underestimate its value and utility.

From colour to contour

RGB may be the foundation, but it's only one part of the picture. The next step is to add shape and depth.

Digital Elevation Models (DEMs) turn flat images into three-dimensional terrain insights. They show us slope, drainage, and erosion risk areas that RGB alone cannot reveal in detail.

In the next chapter we explore DEMs, and learn how understanding "the lay of the land" transforms orchard decisions. From planting to irrigation design, from water use optimisation to orchard performance diagnosis - and everything in between.

SEEING ORCHARDS FROM ABOVE:
WHY RGB DRONE IMAGERY STILL MATTERS

Chapter 4:
Digital Elevation Models

Shaping Strategy with Slope and Elevation

When we think of aerial imagery in agriculture, most minds jump to visuals: colourful RGB images or multispectral maps. But some of the most valuable insights lie in looking at *contour*.

Drone data derived Digital Elevation Models (DEMs) provide unprecedented insights into terrain morphology and land cover. Useful for the modern farmer at both orchard design phase, and orchard monitoring, alongside complimentary drone data metrics.

DEMs are among the most underutilised yet powerful layers available in drone surveying. For perennial tree crops in particular, understanding the terrain beneath the canopy is critical.

Nowadays, drone-derived DEMs provide easy access to visual gradient and slope insights at unmatched resolution. Elevation influences water movement, erosion, soil characteristics, microclimate, and more.

What is a DEM?

A Digital Elevation Model (DEM) is a three-dimensional representation of a landscape's surface.

Generated from drone data, DEMs show us the shape, slope, and height of the land and its features, pixel by pixel. This provides a digital version of "the lay of the land," which can be analysed, and compared over time.

The process involves several steps:

Data collection:
A drone equipped with a high-resolution RGB camera or LiDAR sensor flies over the area in a carefully planned grid pattern.

The overlapping images (or laser returns, in the case of LiDAR) capture every part of the terrain from multiple angles. This overlap is critical - it allows software to reconstruct the surface in three dimensions rather than as a flat picture.

Data processing:
Specialised software then takes the raw imagery and applies photogrammetry or LiDAR processing techniques.

Photogrammetry uses the parallax between overlapping images to calculate depth, while LiDAR measures the time it takes for laser pulses to bounce back from the surface.

Both methods generate a dense "point cloud" that is stitched together into a continuous elevation model.

Output:
The final DEM appears as a greyscale, or colour-coded map, where every pixel carries an elevation value. The lighter or warmer colours most

often indicate higher ground, while darker or cooler tones show lower areas.

- Greyscale maps:
  - Light = higher areas
  - Dark = lower areas

- Colour Maps:
  - Warm tones = higher areas
  - Cool tones = lower areas

Farmers can use these outputs to identify slopes, ridges, valleys, and drainage paths that directly influence orchard layout, water movement, and soil health.

Reading DEMs:
Visual map or numeric model?

Before we go further, a quick note on how DEMs show up in your workflow. A DEM is a map you can see; but it is also a grid of numbers under the hood.

Some platforms lean into the picture-side of representations. Others expose the numbers. You will need to look into which is the right depiction for you.

For example, the *Aerobotics* platform and several others that are in ways similar (like DroneDeploy, Propeller Aero, and the open-source WebODM) can present DEMs qualitatively, with colour-coded layers that reveal relative highs and lows, slope tendencies, drainage paths, and context for orchard decisions. One can read patterns and compare areas at a glance, especially against the other drone data metrics on offer - or across drone survey dates - which becomes increasingly useful. While technical software such as Pix4D, Agisoft Metashape, or QGIS to name a few, can expose the numeric side of DEMs too. Each pixel carries an elevation value.

What this means, and in the context of geospatial data and geographic information systems (GIS), is that a DEM is a representation of a bare-earth terrain surface, and there's a "numeric

Z-coordinate" which refers to the elevation (height) value stored for each location in the model. From those stored values we can quantify slope, aspect, curvature, or run hydrological flow models.

Both approaches (visual and numerical) are valid. The first helps us in acting quickly with clear visual cues. The second lets us measure precisely and is great for modeling scenarios.

In this chapter, we will use the *Aerobotics* view for interpretation and decision support - a smooth entryway into the world of DEMs.

The key idea? To treat the DEM layer as your quick-view terrain lens. Use it to understand the high's and low's; where water moves, and why certain zones persistently underperform.

Whether you read it visually or compute it numerically, the goal stays the same - enhanced planning, smarter interventions, and a clear link between landform and orchard performance.

What DEMs Reveal in the Orchard

RGB and Digital Elevation layers may both come from drone surveys, but the information they provide is somewhat distinct.

RGB shows us the orchard as we would see it with our own eyes: the rows of trees, the colour of the canopy, and the overall layout.

A DEM, on the other hand, strips away the surface appearance and reveals the shape of the land itself and its features. It's a topographic map, showing how terrain could be influencing everything from water flow to soil stability.

A DEM highlights and suggests:

- Where water flows and collects.
- Where erosion is likely, or has already happened.
- Where planting may fail or struggle due to gradient.
- Where machinery access will be difficult (steep gradients, uneven ground).

- Where roads, drains, or terraces would be best positioned.

For example, you might assume water flows from north to south in an orchard because of visible erosion scars in the RGB image. But the DEM may reveal that the true high point is in the north-east, changing your understanding of water movement and your approach to drainage design.

Most of the time when evaluating drone scan results across metrics, it's good practice to note high and low lying areas in an orchard, and to be on the lookout for any topographical - and even vegetative - anomalies or patterns that may assist you with your drone data investigations.

Key Use Cases in Agriculture

Digital Elevation Models unlock terrain-driven insights that often remain hidden in other data layers.

They are not simply topographic visuals; they are diagnostic tools that can shape specific orchard management decisions.

1. **Land use and infrastructure planning:**

   DEMs guide where to place roads, sheds, warehousing facilities, irrigation networks, and worker access routes.

   By aligning infrastructure with slope and runoff patterns, careful planning and execution prevents future erosion damage, repair costs, and inefficiencies that would otherwise be a part of farming operations for decades.

2. **Crop placement optimisation:**

   Most orchards have micro-zones where conditions differ. DEMs help match crops, rootstocks, or cultivars to the right terrain.

   High, well-draining ridges may be necessary to prevent waterlogging and root rot diseases. This practice is essential for most fruit trees and vines requiring sufficient soil aeration and a deep, healthy root system for optimal growth.

   Lower zones might be a better match for more resilient varieties and rootstocks, if drainage is managed appropriately, or where due to soil type and terrain characteristics ridging is not necessary.

   Drone-derived DEM information aids in pointing out these low spots and for guiding in-field sampling and investigation protocols.

3. **Soil erosion control:**

   By modelling slope, curvature, and flow paths, DEMs highlight areas at risk of soil loss before erosion gullies appear.

   This supports decisions on where to terrace, where to plant along contours; or where to run deep ripping lines along the slope to slow water movement and improve infiltration.

   Soil erosion control is especially important in young orchards where bare soil is exposed and highly vulnerable. Acting early prevents gully formation and long-term productivity losses that come with lost topsoil and poor infiltration rates.

4. **Irrigation planning:**

   Because water always follows gravity, DEMs reveal not just where water should flow, but where it will escape to and accumulate.

   Taking note of water movement and pooling areas allows refinement of watering times, emitter placement, pressure management, and orchard zoning.

   Low-lying accumulation areas that might otherwise suffer root stress can be identified and managed proactively with a structured remediation plan.

5. **Drainage management:**

   Persistent waterlogging undermines root and overall plant health, as well as soil structure.

   DEMs pinpoint depressions prone to anaerobic conditions and where plants would be susceptible to root rot; giving farmers evidence to install drainage, re-ridge, or pause development until poor drainage is remedied.

   Oftentimes, in established orchards, a plan to divert run-off is the best way forward. Rather than fight water's natural flow, DEMs are used to guide in-orchard ground works that divert water away from waterlogging-prone areas and elsewhere where it will have less of an impact on plantings.

6. **Precision agriculture and VRA:**

   Terrain strongly influences how inputs move through an orchard. Layering DEMs with multispectral and thermal imagery strengthens tailored Variable Rate Application (VRA) strategies.

   Managing these with gradient and slope in mind ensures fertilisers or pesticides stay where applied, reducing leaching, waste, and environmental pressures while improving uniformity over time.

   DEMs may not only guide VRA strategies to a large extent, but also guide in-field inspections ahead of time so that soil type and structure is taken into account (after sampling) before VRA programs are finalised.

7. **Flood risk and climate readiness:**

   Extreme weather is increasingly frequent, and progressively unpredictable. DEMs highlight flood-prone zones, embankment weaknesses, or areas likely to channel or experience excess water during storms.

   This insight allows farmers to design resilient systems - whether through new drainage networks, precise water diversions, or by shifting infrastructure placement before disaster strikes.

   DEMs are useful layers of data when it comes to planning for inclement and extreme weather conditions. As unpredictable as the weather might be for farmers, there is always room for added preparation (based on reliable data) ahead of unforeseen events.

8. **Stress diagnosis and additional data:**

    DEM layers reveal the “why” behind anomalies flagged by RGB, multispectral, or thermal scans.

    A low-value multispectral zone might trace back to a poorly drained low spot. Low transpiration figures could align with a slope where water runs off too quickly - causing low infiltration rates. These are two small examples, but you get the idea.

    By providing an important piece of terrain-based context, DEMs close the information gap between symptoms and possible causes at the crux of poor orchard performance.

9. **Tracking change over time:**

   As with RGB data already mentioned in Chapter 3; elevation data is not static. Repeated DEM surveys allow farmers to track erosion, sediment buildup, and the effectiveness of corrective interventions year after year.

   Likewise, little to no change in DEM layer information points to terrain stability and confirms known high and low areas so they can continue to be managed appropriately.

   Outside of the scope of ground-level characteristics, follow-up DEM scans are also used to monitor overall canopy heights in terms of a quick visual comparison to older surveys. Used to visually note where tree heights are taller and in which sections of the orchard - this gives farmers a quick look into crop height status compared to surroundings.

All of these DEM interpretation techniques transform this data layer into far more than a static topographic snapshot.

DEMs can be viewed in two ways - as visual terrain maps, or as numeric datasets rich in underlying elevation values. How you experience them depends largely on the platform you're using.

On platforms like *Aerobotics*, DEM layers are presented qualitatively; as colour-coded visualisations that help us to interpret slope, drainage, and relative elevation patterns across an orchard. The focus is on context - seeing how terrain features influence tree performance, stress zones, or water movement at a glance - fantastic for the average user, and easy to grasp and utilise.

In contrast, technical software such as Pix4D, Agisoft Metashape, QGIS, or similar are platforms where added user experience can open the door to advanced analysis of the raw numerical side of a DEMs.

Here, each pixel carries an elevation value that can be analysed mathematically. From this data, slope, aspect, and curvature can be quantified, or hydrological models run to map water flow and erosion risk.

This is an advanced approach, and really shines early on during orchard development, irrigation design, and so forth - utilised by surveyors and planners. However, it can be overkill for established orchards, and this is where I personally prefer the quick and easy *Aerobotics* approach of qualitative visualisations.

Whether visual or numerical, the outcome is the same: DEMs give terrain a voice in the decision-making process. When revisited season after season, they form a historical benchmark - showing how landform and management practices interact over time.

Used together with canopy, multispectral, or thermal layers, DEMs connect the dots between where an issue occurs and why it persists. And as the next case study will show, sometimes a single elevation layer can reveal the root cause of problems that would otherwise remain hidden in plain sight.

Quick Case Study: **Daniel's Erosion Challenge**

Daniel managed a young macadamia orchard planted on a piece of land with a rolling hill. Rolling hills are gentle, undulating landform features that rise and fall gradually, like waves in the ocean.

At first glance, trees in the block seemed uniform and well-established, but within two seasons, certain rows started to lag behind. Some trees were vigorous, while others were stunted and struggling. The variance was puzzling since all trees were the same cultivar, planted at the same time, and managed identically.

A DEM survey revealed the hidden influence of elevation. Subtle slopes were channeling runoff through the weaker rows, while a low-lying corner showed signs of past flooding where fine clay and silt had settled. This altered the soil structure, limited infiltration, and restricted root-zone aeration.

These conditions were preventing oxygen, water, and nutrients from reaching feeder roots. This leads to plant stress, reduced growth, and can eventually kill trees because as we know, roots need oxygen for respiration and nutrient uptake.

The DEM provided the missing context that explained the uneven growth. Multispectral information confirmed suspicions.

Daniel also realised his rows were not perfectly on-contour. However, to get rows perfectly aligned to the contour is rarely practical for blocks already established. In mature orchards he believes in sticking with this simple rule: keep the trees; move the water.

DEM information would guide water-diversion and drainage plans. And armed with this insight, plus experience, Daniel introduced effective corrective measures. He ran deep ripping lines on-contour, added graded contour drains and small strategic swales to spread water; installing cut-off drains to divert peak flows to a safe outlet.

Over the next three seasons, the orchard stabilised. Variance decreased, canopy size improved across the block, and tons-per-hectare figures recovered.

Elevation alone hadn't caused the problem, but without DEM data, the root cause may never have been identified until much later, if ever.

Pairing DEMs with Other Metrics

We touched on this briefly, but to reiterate: Digital Elevation Models rarely deliver their full value in isolation.

Similar to RGB imagery - and quite frankly almost all the other mainstream metrics - the DEMs truly shine when they are layered and compared against additional datasets.

It's in the overlaps where elevation information meets plant stress, canopy size, or transpiration indicators, that the high-value insights begin to emerge.

With RGB:

RGB maps show what the eye can see, while DEMs explain why it looks that way. Erosion scars or standing water in RGB imagery can be validated against slope and contour data in the DEM. This combination prevents misinterpretation and helps separate symptoms from underlying causes.

With NDVI or NDRE:

Multispectral indices highlight zones of stress or poor vigour, but they don't always explain *why* they could be appearing. When those same zones align with elevation features (ridges that dry quickly, hollows that hold water, etc) the diagnosis becomes a lot more obvious. DEMs turn spectral anomalies into actionable and targeted field interventions.

With Thermal or Transpiration information:

Thermal imagery reveals how well trees are transpiring, but terrain often influences water availability. Low-transpiration zones frequently align with shallow and/or sandy soils on slopes, likewise waterlogged land in low-lying areas. Pairing DEM with thermal data distinguishes between likely irrigation faults and natural terrain-driven limitations.

With Canopy and Indicative Vegetative Volume:

Canopy measurements are useful for tracking growth, but when stunted or uneven zones are seen through the lens of elevation, context is added.

A block of smaller trees may not be the result of poor management at all, but the natural outcome of shallow or eroded soils tied to topography. DEMs prevent wasted effort chasing the wrong problem.

This data-metric synergy transforms drone-derived information into a diagnostic toolkit. Instead of looking at each dataset in isolation, DEMs act as the backdrop that grounds all other metrics in the physical reality of the farm. For producers, this means less guesswork, sharper interventions, and enhanced management strategies.

The Cost of Ignoring Elevation

Ignoring terrain is expensive. Farmers who overlook elevation often discover too late that water, soil, and infrastructure have been working against them all along.

Roads wash out because they were designed and placed without due regard for slope.

Drains fail to have the required impact because water doesn't flow where it was expected to flow.

Replanting efforts struggle in low-lying areas that stay waterlogged after every downpour.

Inputs wash away down slopes, forcing higher fertiliser expenditures, or at the very least simply amount to wasted resources.

These problems don't just erode soil - they erode margins. They chip away at capital and confidence, and they turn what should have been long-term assets into recurring liabilities.

A DEM does not guarantee planning perfection, but it removes a lot of guesswork. It shows risks before they become scars on the land or on the balance sheet.

For many farmers, the true cost of ignoring elevation is measured not only in lost yield, but in years of repeating the same mistakes over and over again.

Digital Elevation FAQs

Below are a few Digital Elevation Model questions I've received in recent years, with answers you might find useful - especially if you're new to DEM data layers.

How accurate are DEMs generated from drone data?

With good flight planning and processing, drone-derived DEMs can be accurate to within a few centimetres. Accuracy depends on flight altitude, image overlap, and the use of ground control points (GCPs). For most orchard applications (irrigation planning, drainage, or erosion control) this level of precision is more than sufficient.

How often should I run a DEM survey?

A DEM is not something you need every month, but repeating it every year or two adds value. Annual DEMs allow you to track erosion, sediment buildup, and the success of corrective measures over time. For new developments or high-risk blocks, it's worth running a DEM before or at planting, and then again once trees are established.

Do I need LiDAR, or is RGB enough?

LiDAR can capture elevation under canopy, making it valuable in forestry or dense orchards. But for most open-canopy orchards, RGB-derived DEMs using photogrammetry are more than adequate.

RGB-derived DEMs are cost-effective, easy to capture, and provide enough representative information for the majority of orchard farming analysis and decision making.

Also, it's good to remember that while most drone operators will have a decent RGB sensor payload, they may not have LiDAR equipment - something to keep in mind.

How expensive is a DEM survey?

Costs vary depending on whether you fly yourself, contract a pilot, or use an end-to-end service provider.

In general, a DEM survey is priced similarly to a standard drone imagery flight, and often comes as part of the package with standard orchard mapping surveys. Considering their utility and value in terms of potential long-term savings in infrastructure planning, erosion prevention, and orchard performance diagnosis, DEMs often pay for themselves many times over.

Can DEMs integrate with my other farm data?

Yes. Most platforms allow DEM layers to be overlaid with multispectral, thermal, or canopy data - even soil profiling maps like those for pH, elements, or soil types.

This utility makes elevation insights available as part of a wider diagnostic toolkit. Seeing terrain side-by-side with stress maps or irrigation plans turns DEMs from static maps into tools for action based on reliable insights.

What if my orchard is flat - do I still need a DEM?

Even “flat” land has subtle variations that influence water flow and soil depth. A slight depression can create waterlogging, or result in a frost pocket forming - while a barely noticeable slope can strip nutrients during heavy rains. DEMs reveal these micro-features so they can be managed before they become bigger issues.

Flat land also does not mean uniform tree heights, and DEMs reveal more than surface irregularities, but can also display canopy uniformity in a visual way.

Bringing Elevation into Focus

Digital Elevation Models may not have the glamour of colour maps or the instant appeal of stress indices, but they are among the most powerful tools in the orchard management toolbox.

A DEM is more than a topographic sketch. It's the underlying framework that explains how water moves, where soils accumulate or erode, and why certain rows flourish while others falter.

When viewed in isolation, a DEM provides clarity on slope, contour, and drainage. When paired with other metrics, it transforms into something even more valuable: context.

It turns anomalies in RGB, multispectral, or thermal imagery into stories we can understand. It explains causes, not only symptoms. This context is what allows farmers to act decisively, reducing the guesswork that costs both time and money.

DEMs also offer something few other metrics can. Permanence. Terrain changes slowly, which means most times that a DEM becomes a long-term reference point.

Repeat surveys add even more value, revealing the progress of erosion control, drainage improvements, or replanting efforts. Year by year, it becomes a benchmark for stewardship of both the land and the orchard itself.

In practice, DEMs support a reduction in failures.

They guide drainage channels and irrigation layouts, or prevent replanting into doomed low spots without the necessary interventions made beforehand.

They ensure expensive infrastructure like roads, dams, or buildings are placed where they'll last.

And they provide the confidence that every intervention is structured on the lay of the land itself.

The hidden cost of ignoring elevation is just too high.

With DEMs, farmers gain the ability to see the reality of surface characteristics. To anticipate rather than react, and to make the decisions that respect the terrain as much as the crop.

Elevation matters. And when we understand it well, we farm with clarity, context, and purpose.

From ground to canopy

DEMs give us the shape of the land. The next step is to look at how the trees themselves respond to that land.

Where DEMs show us the ground truth, multispectral imagery (where we're headed in the next chapter) brings the canopy status into sharper focus.

Multispectral values reveal plant stress and vigour in ways the human eye simply cannot.

Together, they connect the health of the soil to the aspects of terrain - addressing the vitality and production potential of the trees themselves.

LOW SPOTS, HIGH STAKES:
DEMS AND THE HIDDEN COST OF ELEVATION BLINDNESS

Chapter 5: **Multispectral Indices**

Detecting Orchard Stress Ahead of Time

Multispectral imaging lets us see what the human eye cannot. It converts invisible signals into actionable insights that guide timely interventions, support healthier plants, and sustain long-term yields.

As we learned in the Chapter on RGB imagery, our vision is limited to a narrow band of the electromagnetic spectrum - this is known as the *visible* spectrum. This is the fairly thin slice of light energy our human eyes can detect, and this is possible due to three types of cone cells that respond broadly to red, blue, and green wavelengths.

Our brains blend these signals into the full palette of colours we recognise each day - but these are in reality only a fraction of the full spectrum. Beyond this narrow window lies a wealth of additional information.

Wavelengths such as near-infrared and red-edge light carry signals about plant stress, nutrient balance, and early disease that remain invisible to our human eyes.

Multispectral sensors capture these bands, transforming them into data layers that can be measured, mapped, and compared over time.

- **Crop health monitoring**: Detecting stress, nutrient deficiencies, or disease *before* symptoms are visually identified in the canopy.

- **Resource optimisation**: Informing irrigation and fertiliser strategies so inputs are applied only where needed and in the correct quantities.

- **Yield maximisation**: Guiding timely interventions that protect production potential - especially so at critical phenological stages where timely interventions produce high-impact results.

- **Environmental monitoring**: Mapping vegetation changes, supporting orchard and habitat assessments, tracking land reclamation projects, and even cover crop management. Used well, multispectral data layers and metrics become multi-use assets.

Multispectral Utility

Multispectral imaging delivers high-resolution data you can collect frequently, making it useful from the individual tree to the whole farm, and from month-to-month checks to year-on-year comparisons - depending on your crop type and your goals.

Multispectral information strengthens data-backed decisions, improves prediction, and saves on resources; supporting targeted interventions with precision.

In essence, multispectral imaging gives us an extra layer of vision - a perspective that extends far beyond visual colour and into the physiological state of the crop itself.

Side note: If any of the terms mentioned anywhere in this book feel unfamiliar, you can refer to the Glossary at the end for further reading.

NDVI and NDRE

Multispectral imagery works by capturing light in both visible and non-visible bands - including near-infrared (NIR) and red-edge wavelengths. These bands are particularly sensitive to changes in plant physiology, often revealing stress long before it can be seen with the naked eye.

While RGB images help us spot obvious differences in canopy colour or structure, multispectral imagery quantifies those differences in measurable ways.

The data is processed into vegetation indices (mathematical combinations of light reflectance values) that correlate with plant vigour, stress, and overall health.

Two of the most widely used indices used on digital platforms for farmers - and the two that we will cover in this chapter - are NDVI and NDRE.

Other Multispectral Indices

Some other indices require red-edge or short-wave infrared wavelengths, which most farm-focused drone sensors don't usually record.

Some of the platform-dependent indices beyond NDVI and NDRE include:

CIred-edge / CInir (Chlorophyll Index) - which estimates chlorophyll content; needs a red-edge (CIre) or suitable near-infrared pairing (CInir).

EVI / EVI2 (Enhanced Vegetation Index) - which reduces soil and atmospheric effects; helpful where NDVI saturates in dense canopies.

GNDVI (Green NDVI) - slightly more sensitive to chlorophyll than NDVI. Useful during fast growth and mid-vegetative stages, when plants are actively photosynthesizing. Often used for intensive crops with overlapping leaves, ie. Maize.

MCARI / MTVI2 - structure- and chlorophyll-oriented indices; mostly for research/advanced workflows.

MSAVI / MSAVI2 (Modified SAVI) - further limits soil influence; handy in young plantings or patchy stands.

NDMI / NDWI - canopy water status proxies.

PRI (Photochemical Reflectance Index) - relates to photosynthetic efficiency; sensitive to short-term stress, band-dependent.

SAVI / OSAVI (Soil-Adjusted; Optimised SAVI) - compensates for exposed soil backgrounds or sparse vegetative cover.

VARI (Visible Atmospherically Resistant Index; RGB) - rough greenness from RGB when multispectral isn't available.

Keeping the before mentioned metrics in mind, I am still confident that NDVI and NDRE do most of the heavy lifting in multispectral orchard analysis.

The additional indices are useful in specific contexts, yet the practical starting point for most farming operations will be NDVI. It is simple to interpret, widely supported on various platforms, and gives a fast read on vegetative presence and vigour - so we will start there.

Let's explore what NDVI measures, where it shines, and how to use it alongside the other metric outputs. Then we'll build toward NDRE for a deeper look into mature, denser canopies.

NDVI: Normalised Difference Vegetation Index

NDVI measures how much near-infrared light (strongly reflected by healthy leaves) is compared to red light (absorbed by chlorophyll).

Formula: (NIR - Red) ÷ (NIR + Red)

The values range on a scale from negative one to positive one (-1.0 to +1.0) and healthy vegetation typically sits above 0.55 NDVI.

In practice, values below zero usually represent non-vegetated surfaces such as water or bare ground. For simplicity, most agricultural data platforms display the index on a scale of 0.0 to 1.0 where the values correspond to the presence and vigour of *living plants*.

While widely used in general fashion for any and every crop for decades now, NDVI is actually best suited for younger trees, vines, cover crops, and plants with smaller or less dense canopies.

NDVI is great for quickly identifying where vegetation is present (or missing) and broadly for health checks, but it's less effective at detecting subtle stress in mature, dense canopies, especially early on before visual signs appear.

From Outer Space to the Orchard

NDVI first became a practical tool in the early 1970s, when Landsat satellites began scanning Earth's surface for the USDA and researchers across the world.

By the early 1980s, scientists were using NDVI to distinguish vegetation types and monitor plant cover and stress over vast landscapes - and we still do this today.

Satellite NDVI made farming visible in a powerful new way - enabling drought monitoring, famine early warning, and crop forecasting across regions and continents.

For generations of farmers, satellite-derived NDVI has been a valuable tool - and platforms like *Aerobotics* and others still provide satellite data to reveal broad patterns across large blocks, sugarcane fields, cash-crops, or inaccessible terrain. During the height of the growing season, when walking the entire orchard or field is impractical, this extra bit of information - albeit low-resolution compared to drone data - comes in handy.

What has changed everything (in a good way) for orchard focused agriculture has been the arrival of accessible UAV technologies and easy to consume drone-derived information.

Today's drones are able to capture the same spectral bands as those captured by satellites, and at much higher resolutions of mere centimetres per pixel - without the need to send anything into orbit to get the show on the road.

Suddenly, in the mid 2010s, NDVI became not just strategic but surgical in its precision. In what seemed like an overnight revolution for early adopters, farmers were able to highlight canopy gaps, identify edge effects, and spatial variability at plant level like never before.

In other words, with the dawn of drone-based multispectral sensor payloads, NDVI had neither been replaced nor had it become less useful. It had become increasingly more precise and progressively agri-relevant.

And while NDVI remains a powerful entry point into stress detection, the specialised NDRE statistics now offer sharper insights into plant health and stress - a topic we'll explore in depth next.

NDRE: Normalised Difference Red Edge

Formula: (NIR - RedEdge) ÷ (NIR + RedEdge)

NDRE uses the red-edge band, which sits between red and near-infrared on the spectrum. Values range from -1.0 to +1.0, and healthy vegetation - for most trees, bushes, and vines - typically sits above 0.44 on the NDRE scale.

A Sharper Lens on Stress

If NDVI was the first step in turning invisible signals into usable crop data, NDRE represents the next evolution.

Where NDVI compares red and near-infrared reflectance, NDRE shifts into the red-edge band - a narrow slice of the spectrum that is highly sensitive to chlorophyll content and photosynthetic activity. Researchers have long recognised the red-edge as a reliable early indicator of plant stress.

Nutrient deficiencies, water imbalance, and disease often cause subtle changes here before they are visible to the eye or picked up by NDVI.

What once required laboratory-grade instruments is now captured by compact multispectral cameras mounted on drones.

This makes NDRE particularly valuable in orchards. In mature, dense canopies where NDVI often flattens out at "healthy" values, NDRE provides a clearer and more nuanced view. It penetrates into the canopy layers to reveal stress signals that NDVI may miss, giving farmers earlier warnings and a chance to act before problems escalate.

NDVI vs NDRE: Which to Use When

Think of NDVI as the broad health check and NDRE as the detailed medical scan. Both indices are valuable, but they serve different purposes depending on crop type, growth stage, and canopy density.

On the *Aerobotics* web platform, which I've mentioned we will lean into for this book, NDVI is used to represent vegetative vigour, and NDRE represents general plant health / stress.

NDVI is best suited for young orchards, vineyards, cover crops, and annuals. It quickly highlights where vegetation is present, how dense it is, and whether blocks are broadly uniform.

NDVI is also useful for detecting obvious canopy gaps or differences in vigour during establishment, when the trees are still small and light penetrates easily to the soil without a blocking canopy mass of real substance.

NDRE, by contrast, comes into its own once trees are well established. Mature canopies that look uniformly healthy in NDVI often hide stress signals that only appear in the red-edge band.

NDRE detects subtle differences in chlorophyll and photosynthetic activity, providing early warnings of nutrient stress, irrigation imbalance, or disease before they are visible on the surface.

For orchards with mixed ages, the most effective strategy is to use both. Apply NDVI to the younger trees, where it accurately reflects growth and canopy development. Use NDRE for the older, denser blocks, where early stress detection makes the biggest difference.

A good analytics platform should allow you to separate age classes into their own "sub-orchard" plant groups, so that NDVI and NDRE are each applied where they work best.

A word of caution: NDRE can underestimate the health of very small plants. Trees with a canopy diameter under one metre will return "low" NDRE readings that don't reflect their true condition. In these cases, NDVI remains the more reliable measure until the canopy expands.

Used together, NDVI and NDRE provide a complementary picture of orchard health - one offering a broad overview, the other revealing hidden detail.

No one multispectral metric is better than the other, it's simply a question of status, context, challenges, and desired outcomes. Used together, these metrics offer the best possible view of vigour and stress in perennial

NDVI Summary:

- Best for young orchards, cover crops, and annuals.
- Highlights canopy presence, gaps, and broad vigour differences.
- Quick to interpret and reliable for small plants.
- Less effective in dense, mature canopies where stress may be hidden.

NDRE Summary:

- Best for mature orchards and dense canopies.
- Detects subtle nutrient, water, pest or disease stress before it's visible.
- Penetrates deeper into canopy layers for a more accurate picture.
- Can underestimate smaller plants (<1m canopy diameter) - take this into account.

Multispectral Metric Guidelines

NDVI:

- 0.00 = Missing or dead plant
- < 0.44 = Stressed, diseased, or in decline
- 0.44 - 0.55 = Healthy (smaller trees) / nominal stress (larger trees)
- > 0.55 = Healthy (both large and small trees)

NDRE:

- 0.00 = Missing or dead plant
- < 0.33 = Stressed, diseased, or in decline
- 0.33 - 0.44 = Nominal stress (larger trees)
- > 0.44 = Healthy (both large and small trees)

Remember that expected or warranted stress will also be picked up with multispectral sensors. We need to remember to take this into consideration. That's one of the reasons why drone scan *timing* is important.

Low NDRE levels may not necessarily be red flags if plant stress is related to current crop load, planned irrigation shut-offs, new leaf flushes, flowering, pruning, and after certain foliar applications.

Timing drone scans with phenology

The value of NDVI and NDRE depends not only on which index you choose, but also on when the survey is flown.

In perennial crops, tree physiology shifts with the seasons. New flushes, flowering, nut or fruit set, and harvest all place different demands on the plant.

Scanning at the right phenological stage ensures the data is interpreted correctly.

Early in the season, NDVI is particularly effective for assessing establishment and canopy growth. Young trees or vines can be monitored for survival rates, gaps, and vigour differences between rows.

As the canopy closes (for deciduous crops, or after heavy pruning of evergreens) NDRE becomes increasingly important, revealing subtle stress clues that NDVI might miss once surface greenness levels off.

During flowering and fruit set, stress detected by multispectral sensors may reflect expected physiological pressure rather than problems in need of correction. A low NDRE value in this phase might simply indicate the tree is allocating resources to reproduction. This is why timing is so important - the same number can mean very different things depending on when the data is captured.

The answer? Conduct and repeat surveys across key stages that build a seasonal story at the right intervals.

- **Post-flowering scans** highlight how well trees recover after bloom. Time these for fruit-set to avoid interference from flowering plants, or better still: after the natural fruit drop stage.

- **Mid-season surveys** reveal nutrient or irrigation-related stress during peak growth (stages like cell division, enlargement, or oil accumulation).

- **Pre-harvest** flights can identify weak blocks, heavy crop loads, or trees that may be underperforming, guiding both yield forecasting and long-term management plans - like post-harvest nutrition programs or pruning protocols.

By aligning scan timing with phenology, farmers avoid false alarms, reduce wasted interventions, and gain a clearer view of true orchard performance.

Multispectral data then becomes less about single snapshots and more about a living record of crop response through the season - expertly timed and executed.

Supercharging Multispectral Results

Multispectral data becomes increasingly powerful when layered with:

- **Digital Elevation Models**:
  Link plant health to slope, drainage, and terrain.
- **Canopy metrics**:
  Relate stress to tree size and growth patterns.
- **Thermal imagery**:
  Assess transpiration, water stress, soil condition, and root health.
- **Crop load and yield maps**:
  Connect stress signals to harvest outcomes.

Quick Case study:
Bruce's Mixed-age Orchard

Bruce managed a mixed-age orchard where older trees were producing well, but younger replacements had recently been planted to fill gaps.

On the surface, the block looked balanced, yet growth patterns varied more than expected.

An NDVI survey brought the younger trees into focus. Several of them showed clear stress, allowing Bruce to pinpoint exactly which rows needed closer inspection. At the same time, NDRE revealed a different story in the mature portion of the orchard.

Although NDVI suggested the canopy was healthy, the red-edge data highlighted subtle but significant stress in certain trees. These fell below the 0.44 NDRE threshold - a warning flag that something was wrong.

In-field scouting confirmed the issue. Irrigation inconsistencies and soil limitations were creating hidden stress that only NDRE could expose.

By marking and syncing the flagged trees to a mobile app, Bruce's team was guided directly to the problem areas for inspection and intervention.

Follow-up surveys showed the value of this approach. With targeted irrigation adjustments and nutrient applications, both the younger and older trees began to recover, and orchard variance steadily decreased.

The combination of NDVI and NDRE gave Bruce the complete picture, helping him manage different age groups with precision and confidence.

Multispectral recap

Multispectral imaging captures light beyond the visible spectrum, revealing crop health insights our eyes cannot detect.

NDVI measures general plant vigour, ideal for younger trees or smaller canopies.

NDRE penetrates deeper into dense canopies, and detects signs of early stress.

Using both indices in the right context provides a more complete picture of orchard health. Threshold guidelines help classify plants as healthy, stressed, or missing, supporting targeted interventions.

Layering multispectral data with thermal imagery, DEMs, and canopy metrics deepens diagnosis and improves decision-making.

Repeated surveys build a historical performance record, enabling trend analysis, intervention tracking, and ROI assessment.

From signals to census

Multispectral indices help us interpret plant health, but the next step is simpler, and no less important.

Before we can measure stress or growth, we need to know exactly how many trees are in the ground, and where they stand.

Plant census statistics give us the numbers behind the orchard, turning a simple headcount into a strategic management tool.

SEE STRESS BEFORE IT STRIKES: HOW MULTISPECTRAL DATA TRANSFORMS ORCHARD MANAGEMENT

Chapter 6: **Plant Census Statistics**

Accounting for Every Tree and Vine

In perennial orchards and vineyards, every plant is an investment. These trees and vines are not seasonal assets that come and go; they are multi-year commitments expected to deliver value for decades.

A macadamia, avocado, citrus tree, or wine vine is a piece of living capital. If we don't know how many we have, or where they are, how can we manage them effectively?

Keeping a precise, up-to-date inventory of these assets is critical for operational planning, budgeting, and long-term profitability. A plant census provides this foundation. It tells us what is in the ground today, what is missing, and how that picture changes over time.

Traditionally, this information was gathered manually. Crews walked row by row with clipboards or tally counters. Field records were updated by hand, often based on estimates or density calculations.

Today, drones and AI have transformed this process. High-resolution imagery and automated plant detection deliver fast, objective, and repeatable results that integrate seamlessly into modern orchard management workflows.

Why Counts Are Important

A reliable plant census supports every layer of farm management. It turns numbers into clarity, helping us align orchard realities with operational, financial, and strategic decisions.

Inventory management

The first and most obvious reason to maintain a census is to know exactly how many productive trees or vines are in the ground. For perennial crops, every plant represents an asset with an expected lifespan of decades. If that number isn't correct, neither are yield forecasts, harvest schedules, or long-term plans. With drone-derived counts, farmers no longer rely on old maps or rough estimates, they can see, block by block, the exact size of the productive asset base they're working with.

Budgeting and replant planning

Replant programmes are expensive, and ordering too many or too few trees has knock-on effects across seasons. A precise census removes the guesswork. By identifying exactly how many gaps exist, and where they are located, we can budget accurately for nursery orders, planting labour, fertiliser allocation, and irrigation adjustments.

This avoids wasted spend and ensures resources go to the right places at the right time.

Insurance and compliance

Insurance claims, compliance checks, and financial reporting all require credible data. A grower who can provide current, high-integrity plant counts has a stronger case when claiming for losses due to storms, droughts, disease, and other unforeseen events.

Similarly, regulators or auditors increasingly expect digital records over hand-written estimates. An accurate plant census not only safeguards claims but builds trust with insurers, lenders, and industry bodies.

Land-use optimisation

Every hectare of land carries a cost. A census helps us understand how much of that land is actively productive and how much is underperforming.

Underutilised corners, irregular planting patterns, or mortality clusters can all be identified and addressed. This might mean replanting to close gaps, rehabilitating poor soils, or even re-purposing sections for windbreaks, biodiversity corridors, a different crop type, or other uses completely.

Optimisation of land need not be reserved exclusively to clearing space and adding trees, sometimes it's about using land more intelligently for the long-term good of the farm - often working with what you already have under management.

Gap identification

Gaps in orchards are more than empty spaces; they are weak links in productivity.

Missing or severely stressed trees reduce uniformity, complicate irrigation and pest management, and drag down block averages - not to mention a cause for waste regarding inputs at block level.

Identifying these gaps or decline in plantings quickly allows us to act before they multiply. In some cases, the solution is simple - a replant. In others, persistent gaps might highlight deeper problems such as soil compaction, drainage issues, or pest and disease hotspots. Census data shines a light on these problem zones, ensuring that we don't just replace trees but also address the underlying causes with quantitative data and objective insights.

The bigger picture

An accurate census doesn't just answer "how many trees?" It strengthens forecasting, budgeting, compliance, and operational efficiency. More importantly, it provides the base layer for all other drone-derived insights.

Canopy analysis, stress detection, thermal data, or precision sampling... they are all of little use (or non-existent at per-plant level) if we don't know which planted assets are where, and how many of them there are. A census, in this sense, is the anchor of precision orchard management.

Return on investment

Ultimately, a plant census is not just about agronomy or record-keeping. It's about protecting and enhancing ROI.

Every tree that fails to produce represents lost potential - whether through lower yield, wasted inputs, or misallocated labour.

By keeping accurate, up-to-date census data, we ensure capital is invested in living assets that deliver returns.

How we can do this involves sharpening budgets, reducing waste, and providing measurable benchmarks that show whether interventions are paying off.

In a business where margins are always under pressure, knowing exactly how many assets are in the ground is one of the most powerful tools for safeguarding profitability.

Can you work out how much each asset cost you each season - the final number? What good are all those calculations if the number of trees or vines you actually have on file are incorrect, or worse, a thumbsuck figure?

Sharpen your ROI calculations and input allocations by knowing your numbers. A good place to start is your plant census, and drone derived data can get you up to speed quicker than ever.

Traditional Approaches

For decades, farmers relied on manual or low-tech methods to track the number of trees or vines in their orchards - but these come with limitations and trade-off's.

These classical approaches have their place in smaller operations but quickly break down as farms scale with time, or when accuracy becomes critical for insurance, budgeting, or replacement planning.

Basic orchard-level extrapolations

One of the fastest methods was to count trees in a representative section of the orchard, then extrapolate those numbers to the entire block. While quick, this approach often produced misleading results. Blocks with uneven survival rates, pest hotspots, or variable soils rarely reflect evenly across the whole orchard. What seemed like an efficient shortcut could create errors that compounded year after year.

Field maps and archival records

Another common approach was to rely on planting records or field maps created at the time of establishment. While these documents provided a useful starting point, they rarely accounted for removals, storm losses, or natural mortality. A record might say 1,000 trees were planted, but after five years, the true figure could be 950 or 900. Without updates, decisions based on these maps were often disconnected from reality.

Labour-based counting

Walking rows with a tally counter offered accuracy at small scales but demanded time and energy that simply doesn't scale in large orchards. Fatigue, human error, and environmental factors (such as poor visibility under dense canopies) all eroded the reliability of physical counts. For farms with tens of thousands of trees or more, this method has become largely impractical.

Planting density calculations

Some producers calculated tree counts based on density formulas - for example, trees per hectare multiplied by total planted area. While useful for quick estimates, this method ignored gaps, uneven growth, or failed replants. On paper the block looked full, but in practice there might be a significant percentage of missing or underperforming trees.

Photography with manned aircraft

Before drones, fixed-wing aircraft or helicopters were oftentimes used to capture imagery for manual interpretation. This was a step forward in perspective but came with high costs and limited frequency. Images still had to be painstakingly reviewed by eye, often leading to slow turnaround times and inconsistent accuracy.

All of these methods shared a common flaw: they could not keep up with the pace of orchard change. Outdated or inaccurate census data led to wasted inputs, insurance mismatches, wayward valuations, and costly rework.

In the modern context of precision agriculture, these limitations have made it clear that a better solution is necessary - and those in the know have already made the shift.

The Modern Solution: Drones and AI

Drone technology has revolutionised plant census work. Instead of relying on estimates, assumptions, or slow manual processes, today's surveys deliver high-resolution imagery and AI-powered analytics in a matter of hours.

Total plant counts and gaps

Every tree is detected, contoured, and logged. Missing plants are automatically flagged, providing an immediate visual of orchard gaps. This creates a true census, not an estimate.

Per-block inventories

The system can break down counts by block, cultivar, or custom designated groupings, offering managers clarity on varietal distribution, age groups, or planting patterns. This is particularly valuable for mixed-crop estates or farms with phased replanting programmes.

Change tracking between surveys

By comparing one survey to the next, farmers can see not only how many trees they have, but also how that number changes over time.

Mortality patterns, establishment success rates, and survival curves become visible, adding a new dimension to farm planning.

When a drone survey measures a reduction in tree counts two or three years after it was planted, it's a sure red-flag moment. When the numbers are stable with little to no change, you can be assured of high survival rates.

Sure enough, one can use the rest of the data metrics to gauge the status of these young trees - but tree census is a highly reliable initial look at orchard status.

Percentage planted

Any good drone data platform will calculate how much of each orchard is truly "in production". Instead of assuming full occupancy, managers can see whether a block is 98% planted, 92%, or even lower (whatever the case may be) and make strategic decisions accordingly.

Perhaps most importantly, drone-based plant detection ties every other UAV-derived metric back to the correct plant - a feature often overlooked or completely missing from zonal-only or layer-only platforms. Keep this in mind when looking at which software platform capabilities you would like to work with.

In addition to obtaining exact plant counts, with per-plant annotations, we get information anchored to actual GPS-tagged trees, ensuring a high level of data integrity when working with the other metrics types.

For example, statistics on what % of the orchard plants are at which NDRE scores. Or which are below or above a certain canopy size.

My advice is to look for per-plant annotation capabilities if at all possible. This way you can get as granular as needed with your data. Think of it as future-proofing your diagnostic efforts. You can always take a step back to zones or layers only if you don't need per-plant granularity for analyses.

Plant status aggregations

For large operations, AI-backed platforms do allow for aggregation. While per-tree precision is invaluable, looking at 20,000 data points on a map can be overwhelming.

Zonal aggregation groups trees with congruent values, simplifying scouting routes, smart-sampling strategies, and reporting.

The result is a census system that is not only accurate but also fast, repeatable, and actionable. It transforms plant counting from a labour intensive task into a strategic management layer that supports several additional data functions.

From budgeting to replanting, to long-term asset planning - the underlying layer of accurate plant figures paired with quick and easy-to-use aggregation tools, puts the power of accurate information back into the hands of production managers.

These capabilities may sound technical to some, but their impact is deeply practical. For many producers, the difference between outdated estimates and AI-powered counts has been the difference between underperformance and profitability.

Lucas, a citrus and macadamia farmer, experienced this first-hand when drone census data revealed a costly gap between what his records showed and what was really in the ground.

Quick Case Study: **Lucas' Missing Production Hectares**

Lucas managed a large citrus and macadamia operation, taking over from the previous manager of a few decades.

On paper, the orchard records looked sound. Planting maps from orchard establishment suggested that every hectare in every block was fully stocked. But when looking at records, yields lagged behind projections year after year, he knew something was probably missing.

The condition of trees in most orchards looked really good as far as he was concerned. There were definitely a few gaps here and there, but surely nothing significant enough to skew yield predictions this much?

A drone-based census quickly told the real story.

High-resolution imagery and AI-backed annotations revealed not only the total number of trees but dozens of small gaps scattered across the blocks.

In one section, more than 18% of the trees were either missing or too weak to be commercially productive. In other areas this figure was consistently averaging between 5% and 12%

When Lucas compared the census data to his original maps and knowledge of the farm, the discrepancy was striking. What he thought was 98% planted land was closer to 82% in the bad section - with no block more than 96% planted.

The shortfall in census counts explained years of underwhelming returns compared to estimates when taking into account planted hectares on file. Armed with accurate numbers, Lucas made some swift changes. He ordered exactly the right number of replants for the gaps - with a few extras for the farm's emergency stock nursery.

He set up a programme to manage and monitor new trees after establishment - and to ensure their survival - and used canopy data from his follow-on drone scans to check that new trees were catching up on canopy size compared to established trees - working steadily to bring down variation figures overall.

Within three seasons, the block's performance had already improved, and the orchard's true productive potential was finally being realised.

Excited about the improvements and then road ahead for these orchards, Lucas was glad he had leveraged the drone data when he did - sooner rather than later. Of course, positive changes were not caused by accurate tree counts alone, but they did form a key part of his management strategy.

Taking a closer look at the data in general helped Lucas to spot other issues and areas suitable for improvement. Since his drone scans did not come with census data alone, but with all the metrics he could get from his service provider, there was always something to learn or optimise.

The census didn't simply count the trees; it changed the way Lucas managed the farm. It took what he knew, added what he didn't, and together with his farming expertise propelled the farm into an era of favourable yields relative to the hectares under management.

Inventory Management in Practice

With accurate census data, management moves from assumption to precision. We can optimise land use effectively by tracking planted versus unplanted hectares. It makes so much sense, but it's traditionally been a big problem to solve for larger and expansive operations and ageing inventory records.

Plan replanting programmes with precision

As with Lucas, knowing exactly how many trees are missing, and where, allows farmers to order the correct number of replacements.

Few farms achieve 100% planting density across every block. Census data shows exactly how much of each hectare is actively productive and where opportunities exist for improvement.

Whether it's filling in gaps, reworking unproductive corners, or reconsidering block layouts, accurate data helps maximise land's contribution to the bottom line.

In other words, a census is not just an agronomic tool. It's also a financial, compliance, and strategic asset.

Knowing the numbers avoids both over-purchasing (which ties up cash flow and nursery space) and under-ordering (which delays block recovery). Replants can also be prioritised by orchard status, performance, or cultivar, ensuring resources are directed where they'll have the greatest return first.

Say goodbye to haphazard planting plans, and hello to strategic decisions backed by reliable census data.

Track cultivar and rootstock distribution

Mixed orchards or phased replanting programmes can quickly have blurred records of what is needed and what is planted where.

A census tagged with cultivar or rootstock information gives us added clarity. This matters for precise orchard management, market forecasting, risk management (for example, cultivar-specific susceptibility to pests or disease), and long-term cultivar planning.

Identify and investigate mortality clusters

Isolated tree losses are inevitable, but clusters hint at a deeper story. Whether caused by soil constraints, drainage issues, frost, or pest outbreaks, these patterns only become obvious when mapped across a census. Identifying them early prevents small localised problems from spreading to become systemic threats.

Countering systemic threats

By definition, countering systemic threats to farming requires a multi-faceted approach centered on building resilience, adopting climate-smart agriculture (CSA) practices, while implementing comprehensive risk management strategies at both farm and policy level.

In the orchard context, a systemic threat is a risk that can move beyond one spot, corner, or row, and can begin to affect the entire block - if not the entire farming operation.

These types of threats emerge when problems are connected and reinforce each other. Water that pools in a low zone leads to root stress; stressed trees become more vulnerable to disease; disease spreads along shared irrigation lines or wet pathways; a local issue grows into a farm-wide problem.

Small failures can compound. Effects are non-linear. You can lose seasons, not only the couple of initial trees you noticed years back.

Clusters are often the first visible sign of an underlying mechanism at work. A "hotspot", line, or epicentre of concern. A few examples include:

- **Hydrology and soil**: Depressions or perched water tables create oxygen-poor root zones. High areas, rocky slopes, or sandy soils low in organic matter struggle to hold water and nutrients.

- **Irrigation distribution**: Pressure losses, blocked lines, blocked emitters, or uniformity issues radiating stress in linear or geometric patterns.

- **Pathogens and pests**: Soil-borne diseases like Phytophthora follow wet feet, insects and disease track weaker trees or spread via pruning wounds, staff, or equipment.

- **Microclimate**: Frost pockets or heat sinks recur in the same terrain patterns each year. Yield suffers, while tree health never seems to recover.

A plant census converts anecdotes into evidence. When gaps and declining trees are pinned to exact coordinates, clusters emerge.

Layer insights with additional data and the picture sharpens further:

- **DEMs** for elevation, slope, and likely flow paths.

- **Multispectral** info for stress signatures around cluster perimeters, and the extent of the stress emanating from these areas on the edge of the dieback zone.

- **Thermal** data for transpiration uniformity, irrigation anomalies, and root health checks.

- **Canopy metrics** to see if the issue is chronic (small, stagnant canopies) or a recent change (a sharp drop from the last survey).

Go from cluster to cause

Work outside-in - validate the pattern, then test your hypotheses. These might look something like this:

- If the cluster sits in a low zone, probe infiltration, drainage, salinity, etc. Frost pockets, stagnant air, reduced light penetration. Compare cluster patterns across metrics *and* across in-field investigations.

- If orchards are on an irrigation system schedule, run a distribution uniformity check and inspect valves, filters, and laterals.

- If stress rings radiate from a point, investigate a point source: chemical spill, pathogen focus, or broken pipe. I've seen these all too often.

- If clustered patterns line up with cold air drainage, review frost protection and row-end escape routes for colder air.

  Work with natural prevailing winds, including anabatic and catabatic movements within the farm's microclimate to direct and improve on airflow.

  Harness winds to your advantage through thoughtful site improvements as needed, or a strategic use of windbreaks.

When a local problem becomes systemic... there are warning signs that you're moving from localised loss to systemic risk:

- The number of affected blocks or irrigation zones increases between drone surveys.
- New clusters appear along high-traffic areas, expected flow lines, or infrastructure links.
- Year-on-year variance widens and replant failures rise in the same patterns as before.
- Remedial actions help briefly, then the problem returns, or improvements stagnate below the desired level.

Catching clusters of missing plants early is about breaking the chain. A timely trench, a repaired valve, a hygiene protocol, or a redesigned exit for cold air can stop a cascade.

That is how a simple census layer helps guard against systemic threats. It keeps local problems local; protects yield stability; and preserves options before the cost curve turns against you.

Fast-track familiarity with census data

Along with RGB data and DEMs, plant counts are arguably one of the most useful first-look metrics when conducting drone surveys - especially so for established blocks.

Where managers are new to a set of orchards, and where they are not as familiar with the farm as they would like... the census data and the rest of the drone survey results are an increasingly useful dataset - especially true in today's farming world.

The reality is that production managers and agronomists move between farming enterprises or manage additional hectares absorbed into the organisation (during business mergers or takeovers) more and more.

It is only inevitable that at some point orchards will change hands, and there is a significant amount of knowledge and experience that is lost during the change-over.

Drone data - including a recent census - offers priceless information readily available for high-impact farming and orchard management from the very first week on the job.

Support insurance claims with objective data

Insurers and auditors increasingly expect evidence, not estimates. Drone-derived census data (time-stamped, geo-referenced, and repeatable) creates an audit trail that stands up to scrutiny for weather damage, disease losses, social unrest, or catastrophic events. The result is clearer documentation, faster processing, and better rapport with underwriters.

This aligns with where modern crop insurance is heading: policies built on measured acres and transparent performance records rather than paperwork alone.

Platforms like *Insure.ag* describe a model where innovative technology, high-resolution imagery, and historical data tailor insurance cover to each farmer and every orchard - where you pay for producing hectares; and where claims are built for farming realities, not boardroom bureaucracy.

That shift pairs naturally with drone-powered workflows. A recent plant census confirms planted versus unplanted areas. Elevation, stress, and canopy layers help distinguish acute event damage from pre-existing issues. Moreover, season-over-season scans provide the historical context adjusters want to see.

Two practical benefits related to insurance:

1. **Eligibility and accuracy:** Verified planted-hectare counts and per-block inventories reduce disputes, tighten sums insured, and help match insurance cover to actual exposure.

2. **Claims velocity and trust:** A clean package (maps, per-tree counts, date-stamped imagery, and notes) supports a friction-reduced process and builds confidence with the insurer's team. It's about digital platform visibility plus support from local agents who act as risk partners - with reliable inventory datasets setting the tone for trust.

In short: pair your census with the rest of your drone layers - and utilise all the accuracy and utility in your insurance plans. You'll protect the operation with credible evidence from day one of your scan; and you'll be ready when the unexpected happens - with data that is both human-readable and machine-verifiable.

Notes on Data Integrity

As you can see more and more... separating the target plant from background vegetation is becoming non-negotiable.

A pixel that belongs to a cover crop or weed is meaningless if it gets counted as part of a tree canopy. Exclusion is just as important as inclusion when it comes to orchard analysis.
Inventory-focused drone scans are no different.

Without precision in identifying each plant, all other metrics risk distortion. Hence my strong recommendation to partner with per-plant level reporting platforms that provide this level of detail.

AI-based plant identification ensures every metric ties back to the right tree or vine. This assures that NDVI stress levels, canopy metrics, or thermal data corresponds to real, living assets in the ground, not a bunch of noise in the dataset.

Data integrity also underpins scouting and sampling. When a weak tree is flagged on a platform, in-field teams need to walk directly to the correct location with ease. Misaligned data wastes time, it undermines trust, and leads to frustration.

Year-on-year comparisons are equally dependent on accuracy. If a tree is mis-identified today, it won't line up with its record in the next survey, breaking the chain of insights. Or worse yet, skewing inventory results, which can be cause for undue panic, and miscalculations.

Integrity may not be the most glamorous part of drone data, but it's the foundation. Without it, even the most advanced analytics lose their value.

Luckily for us, all the best platforms today have data accuracy top of mind and they have the service records and happy customers to back them up.

Census Surveys Best Practices

A plant census delivers the most value when it is part of a consistent and methodical management routine.

Survey annually, plan your dates

At a minimum, conduct a census once a year, ideally after harvest or pruning when visibility is clear, or near the end of the dry season when cover crops are usually absent or have already been cleared or flattened.

Where harvest or seasonal timing is not entirely possible, time your in-field activities like spraying, brush-clearing or cover-crop rolling around the planned drone survey dates.

Leave as much of the area between plants as clear and free as is practically possible. Of course, you will have to juggle in farm activities with phenology, as well as whatever the weather is up to.

A balancing act no doubt, but with careful planning there should be minimal hiccups. Control what you can, and be flexible by a few days either side of your preferred drone scan date - this will help for any unplanned eventualities.

Steps such as these help to provide a reliable year-on-year dataset that supports management, any re-plant planning, production budgeting, and your insurance records.

Capture high-resolution imagery

Low-quality images reduce detection accuracy and create errors in plant identification.

Using capable mapping drones with suitable sensors, flying within the correct parameters, ensures high-integrity data, while every plant is detected, contoured, and annotated accurately and without fuss.

High-quality data not only avoids inaccurate results, it reduces the need for costly reflies and re-mapping. There is also the risk of stalled data processing after collection - pilot uploads flagged for poor quality.

Integrate with smart sampling

A census provides the base map for efficient fieldwork. By linking flagged trees to mobile apps or GPS routes, scouting becomes targeted and efficient. Teams spend less time walking rows and more time addressing actual issues.

Regular surveys transform the census from a static number into a living dataset that tracks orchard change, guides management decisions, and improves profitability over time.

More Than a Headcount

Plant census data is a cornerstone of precision orchard management. By moving from manual counts to drone-enabled AI-processed inventories, farmers gain speed, accuracy, insight, and information that can act on.

Speed = results delivered in days, not weeks or months.

Accuracy = consistent, repeatable, objective counts.

Insight = visibility of gaps, mortality patterns, survival rates, and per-plant characteristics.

Actionability: guidance for replanting, budgeting, yield estimation, and land-use optimisation.

Plant Census Recap

When paired with other drone data layers, census statistics become more than a headcount. They become a strategic tool, showing not just how many plants are in the ground, but how well those plants are performing, how evenly they are distributed, and where management focus should be directed next.

A reliable plant census provides the foundation for every orchard analysis decision. It adds clarity around which assets are in the ground, protects against hidden losses, and links every other data metric back to the correct plant.

By shifting from outdated manual counts to drone-enabled AI surveys, farmers replace guesswork with precision, saving money, strengthening insurance claims, and unlocking opportunities for improved land use for higher returns.

From numbers to nuance

Counting trees is only the first step. What comes next is equally important: understanding how those trees are growing.

In the next chapter, we'll explore canopy metrics - how surface area and indicative vegetative volume metrics reveal the true story of orchard status in the physical sense, and how well these pair with other metrics available today.

PLANT CENSUS STATISTICS:
WHY EVERY TREE AND VINE COUNTS

Chapter 7:
Canopy Area Metrics

Measuring Growth, Managing Variance

When we talk about drone data in the farming sector, multispectral maps and even thermal zoning tends to grab the headlines. Yet some of the most powerful insights come from the most simple of outputs - and canopy metrics fall squarely into that category.

Canopy size may not sound glamorous, but it's among the most practical ways to measure and track orchard performance.

At its core, canopy area refers to the surface area of each tree's foliage as viewed from above. Platforms like *Aerobotics* take this further, calculating not only canopy area (m^2) but indicative canopy volume (m^3).

Together, these figures create a detailed snapshot of structure and uniformity for each plant, and the orchard as a whole.

Why Canopy Data is Important

Canopy metrics are a direct reflection of how our orchards are performing, both tree by tree and block by block.

When we measure canopy area and volume, we are essentially taking the pulse of growth, uniformity, and productivity potential.

This makes canopy data one of the most practical starting points for turning drone imagery into orchard decisions.

Tracking growth

Measuring canopy expansion over time is like keeping a growth chart for your orchard. In young blocks, it confirms whether trees are establishing as planned. In mature blocks, it shows how pruning, nutrition, or irrigation are shaping structure.

A steady increase is generally a positive sign p especially for younger orchards. Stagnation or unplanned canopy shrinkages (pruning, etc) is a red flag.

Smaller-than-expected canopies usually mean lower yield potential, since less leaf area reduces photosynthesis and the plant's ability to support fruit set and filling. Of course, balance matters too. Excessive leaf growth can on the flipside be unhelpful; diverting energy away from fruiting.

Canopy metrics paired with targeted in-field investigations help farmers to really see whether their orchards and vineyards are striking the right balance between vegetative and reproductive growth.

Managing orchard variance

Variance is the hidden tax on orchard profitability. A block with uniform canopies produces a crop that is easier to manage, more predictable to harvest, and easier to market. Wide variance is the opposite. It complicates programmes and labour management, reduces pack-out consistency, inflates costs, and eats into ROI.

Drone canopy data makes this variance visible and quantifiable. Managers can pinpoint struggling rows, identify zones where trees are too vigorous, and plan interventions accordingly - getting down to per-tree level if necessary.

As with plant counts - where we discussed monitoring survival and mortality rates through drone data annotation results, canopy metrics are equally as important. Checking in on canopy size on new plantings as well as plants placed into gaps will provide valuable insights into the success of your efforts and the return on your spend.

Reducing variance is not just a function of agronomy; it's about running a profitable farming business. Uniform orchards translate into fewer unknowns, reduced input costs, increasingly accurate yield estimates, and stronger returns.

Driving precision with VRA

Canopy size is a direct proxy for input demand, making it an essential layer for Variable Rate Applications (VRA).

Instead of treating every tree the same, inputs can be scaled appropriately: smaller trees get a helping hand, while larger trees don't receive unnecessary fertiliser or water.

The same logic applies to pruning: VRA-guided pruning, whether manual or mechanical, can bring oversized trees back in line while avoiding wasted effort on those already balanced.

Farmers who adopt canopy-driven VRA - and tailor their programmes to specific orchard conditions and needs - consistently report reduced costs in the longterm, less waste of resources, and healthier orchards.

Timing canopy surveys

To make canopy data meaningful, timing is everything. The best practice is to fly at consistent stages each year:

- **Early season**: after flowering and fruit set (ideally after natural drop) to assess vegetative growth potential.

- **Mid-season**: at the start of cell enlargement or oil accumulation, when structure is building.

- **Post-harvest**: before end-of-season pruning, to gauge canopy fullness and orchard variance.

Avoid drone scans during flowering (unless specifically measuring bloom intensity) or immediately after pruning (unless specifically to monitor pruning efficacy or labour competency). In short, both flowering and pruning distort results for classic canopy focused surveys.

A single well-timed drone scan already provides valuable insights. Adding a second within the season deepens the picture, especially if tied to a specific management protocol. A third survey is often possible - even needed - but diminishing ROI and practical constraints related to time and resources for larger operations or those short on staff mean that a tertiary scan is rarely necessary.

As a side note here, it's important to realise that shorter growing windows for deciduous crops like grape vines, stone-fruit, pome-fruit, pecans, etc... means that two or three drone scans would need to fit into a smaller portion of the calendar year compared to evergreen crops like citrus, avocados, mangos, macadamias, and so forth.

In some instances your available window for drone scans might just depend on your growing conditions and microclimate, or even the specific cultivar selection of your target crop - like with pomegranates, which may or may not lose leaves in winter.

Either way, keep growing season length and your growing environment in mind when it comes to drone survey timing.

Quick Case Study: **Abigail's Mixed Orchard Analysis**

Abigail managed a mixed-crop farming operation of avocado, mango, litchi, and citrus - where her team relied heavily on visual inspections.

From the ground, her trees looked healthy across the board, yet yields kept falling short in certain orchards. A drone canopy survey revealed why.

The analysis showed significant variance in canopy sizes: some trees were thriving, others lagging behind. These were hard to quantify at a glance since they were mostly sprinkled here and there, with only some instances occurring in straight rows, or in patches clustered together.

Further investigations using her drone data results pointed to two culprits: uneven irrigation distribution (water delivery) and shallow soils over underlying bedrock. Shallow soils can be defined as those with a root-restrictive layer found within about 50cm (less than 20-inches) of depth. These conditions present very specific challenges for orchard agriculture.

By adjusting irrigation zones and ironing out system problems, the improvement plan was off to a good start. Next, employing targeted and intensive soil management regimes - specifically those around water and nutrition management - meant Abigail addressed the majority of her problems. Both areas of improvement were monitored, tweaked, and improved. The following season's survey showed clear results: reduced variance, healthier growth, and a more predictable crop.

With improved clarity, Abigail not only boosted yields but also sharpened her budgeting. Resources could now be directed where they were truly needed (and with higher clarity on costs), Abigail, her consultants, and staff were empowered by the insights, and she landed up protecting ROI. Each drone survey gives her the confidence to not only plan ahead but action those plans.

Canopy Data in Context

Canopy data rarely stands alone. And as with all the metrics we have covered ahead of this chapter, its value multiplies when it's layered with additional datasets available today:

- **RGB imagery** validates observations with visible context and the not-to-be-discounted view from above.

- **DEM layers** explain growth differences driven by slope, drainage, or soil depth.

- **Multispectral indices** clarify whether smaller canopies are simply young or genuinely stressed and lagging behind.

- **Plant census statistics** pair perfectly with canopy metrics to provide a holistic view on orchard status, survival rates, and the level of progress after planned interventions.

- **Thermal data** links canopy performance to transpiration efficiency, root health, irrigation effectiveness, and adds additional context to multispectral insights.

When integrated, drone-derived datasets turn canopy metrics from a descriptive snapshot into a diagnostic toolkit. This layered approach supports better decisions across irrigation, nutrition, replanting, and pruning strategies.

From canopy sizes to water-use insights

Canopy metrics may not grab headlines, but they are indispensable. They track growth, highlight variance, and underpin VRA decisions.

With carefully timed surveys, canopy data builds a performance record that connects management practices to outcomes.

Every reduction in variance reduces waste, strengthens predictability, and boosts ROI at harvest.

But canopy size only tells part of the story. The next question is: how efficiently are those canopies using water to sustain their growth?

In the following chapter, we'll turn to **thermal metrics**, which reveal transpiration patterns, water use efficiency, and hidden irrigation gaps that canopy data alone cannot explain.

MANAGING GROWTH AND VARIANCE:
THE DRONE ADVANTAGE WITH CANOPY METRICS

Chapter 8: **Thermal Metrics**

Representative Heat Maps and Water Use Efficiency

Drone-based thermal imagery gives us a powerful, non-invasive way to understand how well orchard assets are utilising water.

By visualising differences in canopy temperature, we can identify early on whether trees are transpiring efficiently or showing signs of imbalance, even before visual symptoms appear.

Healthy trees regulate temperature naturally. Through transpiration, water evaporates from the leaf surface, cooling the canopy and maintaining internal balance.

When this process slows down - due to water stress, soil compaction, disease, or root damage - the affected trees "warm up".

The contrast between these "hot" and "cool" areas in an orchard or vineyard setting creates an impressive visual map of orchard health and performance at the time of surveying.

It's important to understand that most drone-based thermal imaging for agriculture does not measure absolute temperature in the same way as a thermometer would. Instead, it highlights *relative* differences.

These maps are representative, not entirely diagnostic on their own, but reveal patterns and contrasts that direct attention where it's needed most.

In precision agriculture, it's not the exact degree that matters, but the *variance* in transpiration that highlights issues otherwise unseen or overlooked.

The real advantage with thermal-based information is that we are gaining an additional and unique perspective from above.

Similar to Multispectral information already discussed, one is able to pick up on unseen values from the available spectrum we cannot see with the human eye. This time though, it's less about light, and more about heat.

Heat is the transfer of thermal energy from the kinetic motion of molecules, while visible light is a form of electromagnetic radiation that we as humans can see.

Though distinct, light and heat are related because hot objects emit infrared radiation, a form of invisible light that we feel as heat.

From Heat to Hydration: What Thermal Data Tells Us

Every pixel in a thermal map corresponds to how efficiently a plant exchanges water with its environment.

When viewed across hundreds or thousands of trees, these differences become clear indicators of how uniformly an orchard is functioning. The key is in how to identify the patterns - similar to the clustering and zonal examples we covered in previous chapters.

At the heart of the current focus on thermal based analysis, there are two key metrics I will focus on for the most part: Transpiration Uniformity **(TU)** and Transpiration Variance **(TV)**.

Light versus heat

Thermal imaging on drones measures the infrared energy emitted by surfaces. In orchards this reveals relative differences in canopy temperature between trees, rows, and zones.

Healthy, well-watered trees with healthy root systems and soil conditions tend to run cooler; stressed trees often appear warmer in comparison. That pattern is a fast proxy for transpiration behaviour across a block.

On platforms like *Aerobotics*, thermal layers are presented as representative maps that highlight contrasts rather than absolute temperatures.

Thus, the emphasis is on Transpiration Uniformity (TU) and Transpiration Variance (TV) - practical ways to see how evenly water is being used, to flag irrigation issues, compacted soils, or problem microclimates.

Some radiometric cameras and specialist software do allow pixel-level temperature export and more advanced modelling. If your workflow needs that, treat it as an advanced path; the operational value for most teams still comes from the relative pattern and how quickly these representations can direct in-field checks.

Making progress on your data results is about efficiency. When it comes to thermal based info, avoid getting lost in a lengthy deliberation of the finer details - all of which will only slow you down.

Thermal Data Capture Tips

Thermal scans are highly sensitive to timing and environmental conditions. To capture meaningful data, flights should be planned when temperature contrasts between healthy and water-stressed trees are most pronounced.

Time of day

The best results come from mid-morning flights, once the canopy has warmed enough to differentiate transpiration behaviour but before midday heat evens everything out.

Early-morning flights can over-emphasise cold air pooling, while late-afternoon flights often produce flattened contrast as canopies cool again.

Keep goals in mind when arranging flight times:

Are you focused on layered information (general orthomosaic outputs) or per plant or zonal outputs? If you're interested in the trees or vines instead of generalised areas, aim for as close to solar noon as you can manage.

If you plan to monitor specific irrigation system valves or zones, decide on drone flight times according to irrigation scheduling.

The time after an irrigation zone has been watered could be a good window into which you could get a look at transpiration uniformity, possible drainage issues, pressure problems, and so forth. Likewise for the time right after some rains - if conditions and weather allow.

Weather conditions

Thermal imagery depends on stable sunlight and minimal wind. Avoid flying after rain (unless for specific reasons like checking on drainage - but even then - wait a few hours) during cloud cover, or in gusty conditions.

Moisture on leaves or strong airflow can distort canopy temperature readings and mask subtle differences - foliar sprays whether ground-based or aerial, likewise.

Seasonal timing

Thermal data is most valuable during periods of high evaporative demand; typically late spring through midsummer. This is when irrigation systems are under pressure and water-use efficiency becomes critical.

For crops like macadamias, avocados, or citrus, running thermal surveys just before and during key fruit-filling stages provides the clearest picture of hydraulic performance and irrigation effectiveness.

Consistency

As with all drone metrics, consistency builds insight.

Conducting thermal scans at the same growth stage each year, under similar environmental conditions, ensures results are comparable and meaningful.

Thermal surveys, when timed correctly, can transform routine irrigation checks into high-value diagnostics - revealing inefficiencies before they turn into yield losses.

Repeat scans at similar times and phenological stages for like-for-like comparisons. This is key, as the thermal sensors on drones today are extremely sophisticated and sensitive - if drone scan timing is off or conditions are too dissimilar compared to your previous scan, the information may not be as comparable as you would like.

What you *can* do across surveys though, is to gauge for long-standing issues by comparing transpiration patterns with NDRE values from current and previous drone flights to get a clear view of how long a certain issue has been around, or to gauge the spread and scope of a problem.

Context layers

Pair thermal with DEM layers (drainage and slope), NDVI/NDRE (stress depth depictions), canopy metrics, census insights, and any in-field telemetry like weather stations, localised atmospheric conditions, soil moisture probes, and the like. Use all the insights you can to separate cause from symptom.

Uniformity vs Variance

With visual thermal interpretations it's all about patterns and contrast. Use TU and TV information together for best results when it comes to deciding on how to act on your drone scan results.

Transpiration Uniformity (TU)

TU measures how consistently trees are transpiring.

A high TU value suggests most trees are cooling themselves at a similar rate - meaning irrigation, root health, and soil conditions are relatively balanced.

A low TU score, by contrast, indicates variability. Some trees may be shutting down, others over-transpiring.

These patterns often point to uneven irrigation, waterlogging, or root damage.

Transpiration Variance (TV)

TV quantifies the degree of variation in transpiration between individual trees.

It's one of the clearest indicators of orchard balance or imbalance.

A low variance on the *Aerobotics* platform for example (< 0.25) means uniform water use.

Moderate variance (0.25 - 0.35) signals emerging irregularities that should be inspected.

High variance (> 0.35) means immediate attention is required; where water movement, soil structure, or system calibration are likely compromised.

Quick-reference TV Interpretations

- **< 0.25** = Low variance; good, low concern. Monitor only extreme outliers.

- **0.25 - 0.35** = Medium variance; investigate zones for inconsistent irrigation or soil conditions.

- **0.35** = High variance; priority inspection required. Target blocks for field checks immediately.

A high variance score isn't a temperature reading - it's a signal. It tells us that something is disrupting the natural equilibrium of water flow within the orchard system.

Understanding Turgor Pressure

Turgor pressure is the invisible force behind a plant's rigidity and vitality. Basically, when a plant loses water faster than its roots can take it up, the volume of water in the cell's central vacuole decreases. This reduces the turgor pressure.

When turgor pressure drops, cells collapse slightly, causing wilting - even if the tree is still alive. And so turgor pressure is crucial for maintaining a plant's structure.

Thermal imagery indirectly reflects this physiological process. Trees with strong turgor pressure stay cooler, as water moves freely and evaporates efficiently through open stomata.

Trees with low turgor pressure, however, struggle to regulate canopy temperatures, appearing warmer on thermal scans. By monitoring these patterns with sensitive drone-backed equipment, farmers can identify where physiological water balance is compromised; well before visible wilting or leaf drop occurs - and for short term events, even before any NDRE alarm bells have started to ring out about plant stress.

Transpiration and monitoring stress

In order of earliest to latest, here are the metrics that provide feedback around plant stress:

- Thermal data (low transpiration)
- Multispectral data (NDRE < 0.33)
- Multispectral data (NDVI < 0.44)
- Canopy metrics (shrinking trees / no growth)
- RGB visuals (yellowing, visual decline)
- Plant census results (trees already lost)

Quick Case Study:
Henry's Hidden Variance

Henry managed a mature citrus orchard that, at first glance, appeared uniform. Canopy size, NDVI, and NDRE values all suggested good health according to guidelines.

However, yields from one section of the farm kept slipping, despite consistent management practices.

A thermal scan shifted his perspective. The block's Transpiration Variance (TV) measured 0.39 - above the normal range. While visually similar in canopy size, thermal maps showed certain rows consistently warmer than others.

Field checks revealed a pressure drop in the irrigation mainline feeding that area. Some emitters were barely dripping, while on the other side of the orchard they ran at full flow.

The fix was simple but significant: new pressure regulators and line balancing, together with flushed lines to ensure no obstructions in the lines on the lower portions of the orchard.

Follow-up scans showed TU improving across the block, with variance at new lows. The next season's yield variance dropped by 12%, coming in close to the rest of the farm's performance.

The insight into transpiration imbalances saved Henry from unnecessary nutrient applications and unanswered questions; with redirected investment toward irrigation improvements - a clear ROI win with positive knock-on effects that will be felt for seasons to come.

Practical Applications in Orchard Management

Thermal data has a wide range of uses across the farming calendar. To summarise, here are a few more instances where it adds value:

- **Irrigation performance:** Detects inconsistent water delivery, pressure issues, faulty emitters, or blocked lines.

- **Water use efficiency:** Identify where trees are over or under-transpiring and optimise irrigation schedules accordingly.

- **Soil condition assessment:** Highlight compacted or waterlogged areas where poor aeration limits root function.

- **Frost-risk mapping:** Spot cold air pooling zones at dawn to plan protection strategies.

- **Soil moisture probe placement:** Use TU maps to identify representative or problematic zones for soil sensor installation.

- **Variable rate irrigation (VRI):** Combine thermal data with DEMs and canopy metrics to fine-tune irrigation delivery across slopes or heterogeneous soils.

Thermal imagery adds a physiological layer to data-driven farming. It doesn't replace multispectral or canopy insights; it completes them.

Layering Thermal with Additional Data Metrics

You probably saw this section of the chapter coming, didn't you?

As with the previous metrics - and for you to make the most of your thermal data - it should be analysed alongside the other metrics for maximum effect.

And perhaps more so than any other metric, because it reveals a side of the canopy status that could potentially not yet be seen in any other way - other than looking at canopy surface temperatures between trees or vines, and across whole planting areas.

- **With RGB imagery:** Add visible context to locate the affected areas accurately, or validate/invalidate initial assumptions.

- **With DEMs:** Understand whether topographical characteristics; drainage, pooling, or slope are influencing transpiration.

- **With Canopy metrics:** Link canopy size and shape to cooling efficiency. Sames sized trees transpiring differently? Time to look a little closer.

- **With Multispectral indices:** Confirm whether temperature anomalies align with nutrient, pest, or disease-related stress.

Together, these layers form a complete diagnostic system, where each dataset reinforces the other. Combined with farming experience, in-field telemetry, and a plan well executed - it's powerful stuff.

The Value of Consistency

When tracked over time, thermal metrics reveal how orchard interventions perform. An improving uniformity score (TU) shows irrigation and root-zone health are stabilising.

Rising variance figures (TV) indicates worsening imbalance and the need for review - especially where NDRE and canopy variance figures are growing too.

The ability to quantify this change in uniformity season after season transforms water use from a guessing game into a measurable management process.

From Heat Maps to Actionable Insight

Thermal imagery translates heat into understanding. It shows us how efficiently trees use water, how evenly orchards perform, and where intervention is most urgent.

TU and TV make irrigation performance measurable - giving us the power to fine-tune operations and prevent silent yield losses.

When integrated with other metrics, thermal data bridges the gap between plant physiology and management practice. It helps us see not just how the orchard looks, but how it functions.

From Heat to Harvest

Having explored how thermal data reveals water use efficiency and physiological balance, we now turn to **Precision Sampling** - the next step in turning aerial insights into smart ground-truth actions.

In the following chapter, we'll see how AI-driven sampling transforms data into decisions, ensuring that every inspection counts and every datapoint drives progress.

COOL TREES, HOT PROBLEMS:
THE THERMAL METRIC EVERY FARMER SHOULD TRACK

Chapter 9: **Precision Sampling**

Smarter Sampling, Reliable Data, Better Decisions

Sampling has always been the bridge between what we see and what we know. In orchards and vineyards, it is how we convert observations into evidence that guides irrigation plans, nutrition programs, crop protection, and harvest logistics.

The problem has never been the *idea* of sampling. The problem has been *where* and *how* we choose to sample - and whether those samples truly *represent* the status of agronomic assets.

Drone data changes the game.

When aerial metrics are combined with artificial intelligence and machine learning models, sampling shifts from guesswork to guided strategy.

We stop wandering across orchards (or stop choosing the easy spots to sample) and we start to target and track; guided by protocols backed with data.

Every visit to the orchard has a reason, and every datapoint won moves you toward a calculated action.

From Guesswork to Guidance

Traditional sampling often relied on habit.

"Walk every fifth row."

"Pick a tree at random, make sure it looks like the average tree."

"Sample the first ten vines in the second row near the entrance, that will be fine."

"Do a diagonal sweep across the orchard from both corners, sample every few trees."

It may have felt fair enough in the past, and it was simple to do. The catch is that orchards are not uniform, and humans are prone to biases and inclinations of every kind.

With slope, soil depth, drainage, cultivar, rootstock, age, and management histories that all vary within a few hectares or acres, randomness is a pretty weak proxy for fair representation.

Unless your data volume is something extraordinary, you are going to come up short on representing the whole group of plantings each and every time.

Although you might strike it lucky every now and then... as the old saying goes, "Even a broken clock is right twice a day".

Precision sampling begins with a different starting point. Not “which trees are closest”, or how few trees can we get away with for our sampling, but “which trees together represent the full range of performance and conditions in this block.”

This is a subtle shift with enormous consequences. It means we are measuring the orchard we actually have, not the orchard we imagine we have.

The Breakthrough: AI-Guided, Representative Sampling

Modern platforms ingest the layers you have already seen in previous chapters. RGB for context, DEM for terrain, multispectral indices for signals of stress, canopy metrics for structure status, thermal for water use efficiency, census data for plant identity and management.

And then AI comes in and does three things that are hard to do reliably by hand, and at scale.

First, it stratifies. The platform divides the orchard into zones and clusters that capture true variability. Think of it as an automated version of stratified sampling that is tuned by spatial patterns, not arbitrary grids. Areas that are similar are grouped. Areas that are different are separated.

Second, it balances. Within each stratum, the system selects sets of trees that together represent the distribution of conditions you care about. You get a deliberate mix of typical performers and outliers. Not only the mean, but the tails. This improves your ability to diagnose causes, predict outcomes, and plan for the future.

Third, it optimises logistics. Waypoints are sequenced into walkable or drivable routes. The order reduces backtracking, respects terrain and access patterns, and groups tasks by worker or team. Sampling stops being a maze and becomes a route. An answer to chaos.

Why this approach beats "random"

Random sampling can miss the very patterns that drive cost and risk. It under-samples small but critical problem areas, over-samples easy-to-reach rows, and produces year-to-year datasets that are hard to compare because the randomness changes each time.

Moreover, it's hard to replicate, and creates flawed protocols. How can we expect all workers to "randomly" sample in the same way across an estate? It's just not sensible.

AI-guided sample waypoint selection gives you repeatable representativeness - the objective advantage.

The logic under the hood.

Although platforms each have their own recipe, representative selection commonly uses a blend of spatial stratification, feature (metric) clustering, and diversity maximisation.

In practice this means the algorithm considers canopy area, NDVI, NDRE, thermal variance, elevation, contours, distance to edges or roads, and sometimes historical change, and even mixed plantings (more than a single cultivar inan orchard).

The goal is to choose trees that are far enough apart in the feature space to be well spread, yet close enough in physical space to be practical to visit. Due to the nature of farming and variability in blocks, striking this balance is not always possible, but more a guiding light.

Algorithms respect practical constraints. You can set a sampling budget, cap trees per zone, exclude inaccessible rows, and digitally save locations for key trees that you want to sample repeatedly.

Designing a Precision Sampling Plan

Before we get into specific methods, it helps to sketch the plan that will hold the concepts together in practice.

Define the decision: Are you validating a nutrition tweak, chasing a suspected irrigation fault, creating yield estimates, or diagnosing a puzzling patch of decline?

Clarity on the decision forces clarity on the metrics and on the number of samples you truly need.

Choose your layers: Use canopy size to map structure, NDRE for subtle physiological shifts, thermal TU and TV for water use efficiency, and DEM for slope and pooling. RGB gives orientation and context so teams know exactly what they are walking into (or where they are on the map).

Pick a sampling angle: Use zonal when you want good orchard representation. Use per-tree outliers when you want to diagnose specific problems. Go the cluster-based route when estates are large and you need scale without losing signal. Go for historical change detection when the question is whether interventions are working or not.

Set sizes and standards: Agree on counts per zone, per cluster, or per cultivar group - have consensus on what data is captured at each tree and when. Fruit counts, diameter measurements, leaf-based or soil-focused tests, photo angles and types, required notes, etc.

Clear communication on what is expected and required for data collection keeps your teams on the same page, their actions consistent, and datasets comparable.

Schedule for phenology: Scan and sample at stages that match the mission. There is little value sampling fruit for size too early in the season, or checking water use on a cold, still day that suppresses contrast.

Good timing sharpens what the data reveals and the patterns become a lot more obvious.

Zonal Sampling: Represent the Orchard You Actually Have

Zonal sampling divides an orchard into performance zones that capture real differences.

The most common approach is to build zones from NDRE and canopy size, check them against DEM and thermal, then label them as high, medium, and low performance, or by any segregation method that fits your operation.

Sampling from each zone keeps the dataset balanced. You are not guessing where to look. The zones tell you, and you cut away the guesswork.

What to measure

In high-performing zones confirm targets and look for excess vigour that may hurt fruit counts or quality. Timely pruning to control vigour might be required, or you may need to watch out for crop-load stress if it's simply "a very good year" with an over abundance.

In medium zones you search for constraints that are holding trees just below potential. Look out for the quick wins, or if everything seems fine, move on swiftly.

In low zones you try to separate transient issues from structural ones. Thermal can reveal whether water delivery or root function is the primary limiter. DEM can confirm thermal figures, or reveal slope and pooling problems that undermine crop performance.

How many sampling waypoints?

A practical starting point is 3 to 5 trees per zone for fruit-level metrics in perennial crops - with 3 to 5 zones per orchard. Easy numbers to remember.

You can adjust this protocol for block size and variability as needed. If a zone is highly variable inside itself, split it further or add a small per-tree outlier set to the plan.

If doing research work, you can bump this figure up to 15-25 trees per zone in 3 to 5 zones (multispectral data) or anything from 3 to 20 zones for canopy metrics (depending on the platform).

Transpiration uniformity information lends itself well to a 4 zone strategy - a classic method used to sample thermal outputs - an approach taken from the days of manned aircraft mounted thermal sensors.

The amount of sample waypoint locations will depend on your capacity, orchard and zone shape/s. Whatever you decide, keep the approach standard and make it repeatable to reduce data collection variability and human error, like subjective sampling and biases.

Quick Case study: **Pablo's Citrus Zones**

Pablo managed around 56 hectares (about 138 acres) where fruit size skewed small in parts of the orchard. NDRE Zonal maps separated the orchard into three performance bands.

Sampling 5 trees per zone for fruit count and size gave a realistic outlook that tightened assumptions, harvest logistics and go-to-market (GTM) planning.

Cross-checking with thermal datasets and fruit size sampling (*Aerobotics TrueFruit* spatial map overlays) showed that zones with smaller fruit happened to show up warmer on hot afternoons.

A targeted irrigation adjustment and water pressure balancing brought the zone into line for the remainder of the season. The value was not only in the fix, but in the confidence in the sampling methods and the forecasting tools drone data and AI-backed apps brought to the party.

Zonal information and ROI

Zones turn variability into a manageable structure. You can budget inputs per zone, sample per zone, harvest per zone, set expectations per zone, and brief teams per zone. With easy to understand visuals on mobile apps and web platforms - even print-out's and PDF's - anyone on the team can utilise these representations with ease.

Over time, you want the low zones to shrink (suboptimal values) and the high zones (your optimal values for NDRE or NDVI for example) to grow.

In the case of canopy metrics, a uniform block will be one solid zone colour or gradient depicting near congruent canopy structure values across the orchard.

In the case of thermal data, you may have the same amount of zones on your platform display, but with less contrast and colour or gradients that seem to fade into one another as opposed to something high contrast.

Per-Tree Outlier Sampling: Diagnose What the Averages Hide

Outliers carry information. A handful of weak trees can signal an irrigation valve fault or a soil borne issue.

A handful of oversized, dark-green canopies can signal over-vigour that could hurt fruit quality or flowering potential.

Per-tree outlier sampling uses AI to pinpoint the individuals that deviate most from the cohort on metrics that matter. You then visit those exact trees.

What to look out for

For weak outliers, pair NDRE with thermal. If NDRE is low and transpiration is low, water use is likely the limiting factor. Low availability for the canopy due to dry or waterlogged conditions.

If NDRE is low and transpiration is normal, nutrition or disease may be the driver.

Outlier monitoring protocols

At each outlier tree, follow a repeatable micro-workflow.

A set of photos for context. Fruit counts, and AI-app diameter measurements. Leaf or soil samples if the situation calls for it. A short note of what you see and what you changed - if anything.

Precision is not only *where* you go (precise objective locations). It's also *how* you collect your data, and *what kinds* of data you collect when you get there.

Quick Case Study: **Lucas's Hidden Compaction Line**

Lucas struggled with a row of underperforming avocados that kept slipping under the radar during sporadic checks.

Outlier sampling flagged a string of trees with smaller canopies, low indicative canopy columns, low NDRE values, and lower than average transpiration figures - all in a straight line between two access roads either side of the block

A penetrometer during the targeted in-field inspection told the rest of the story. Compaction from repeated traffic passing through a "shortcut" between access roads had created a root-limiting layer.

Deep ripping on the traffic line, with some compost mulching and a root health programme - plus a change in access for machinery on this route - brought those trees back within a season.

The lesson was a simple one: averages hid the issues, while outlier identification revealed the patterns needed for action, operational changes, and orchard improvements.

Cluster-Based Sampling: Scale Without Losing Signal

Large estates need scale. Cluster-based sampling uses AI to group similar trees in the same area into clusters based on canopy, NDVI or NDRE, thermal behaviour, or elevation context.

You then sample a small, consistent number from each cluster. This keeps field time reasonable while maintaining statistical coverage of the orchard's diversity.

How this differs from cluster sampling mentioned in previous chapters is that in the case of drone data metrics like NDRE, canopy info, and transpiration uniformity, "clusters" are patterns of individual trees that form patterns and shapes denoting possible underlying causes and limiting factors. They hint at scope and prevalence, and help to guide next step decision making.

AI sampling "clusters" are groups of trees chosen as the source location for a single data collection made in the said location. Handy in the event of too few fruit on single trees (yield estimation sampling) or where the sample location is based on zonal maps - where a single sampled tree might not be as representative as you would like.

Where the precision comes in is in the location identification of where you will cluster the trees or vines to sample, and in your decided collection protocol which needs to be repeatable. For example: 2 trees sampled either side of the location pin placed in the inter-row, where the waypoint pin is in the middle of an NDRE zone.

Cluster sampling design choices

You can cluster by a single dominant feature when the question is narrow, such as canopy size during a pruning audit. Or you can cluster on a blended feature set when the goal is to represent multi-factor variability.

The platform will typically expose or suggest the number of clusters, their sizes, and their average metrics so you can sense check before you walk.

Desired turnaround times and labour capacity is also a factor you should keep in mind. A large data volume is favourable, but data collection decisions need to be practical and sustainable relative to your goals and available resources.

Quick Case Study:
Melanie's Macadamia Estate

Melanie manages 217 hectares (about 536 acres). To get her crop yield estimates as accurate as possible for her own records and for her processing partner, her team decided to use their drone scan data to visit 5 clusters per block, 4 trees per cluster.

Nut count sampling commenced after natural nut drop, and was anchored on canopy and NDRE information for most orchards. Transpiration zones were used as a tertiary data source for her orchards with high NDRE and canopy size uniformity figures.

Sampling two trees per cluster gave a reliable read on fruit set and nut fill, while keeping the sampling workload manageable. With less confusion, and precision guided sampling waypoints, field time dropped by roughly half, while their yield forecasts tightened dramatically, aligning with the reality on the ground.

Historical Change Detection

The best question in farming is not only "what is happening," but "is it getting better."

Historical change detection compares surveys across seasons and flags where the trend matters.

Growth that accelerated after a drainage fix. NDRE that recovered after a nutrition change. Thermal uniformity that improved after pressure balancing.

Smart and precise repeat-sampling done in drone-data based target areas validate any trends with on-the-ground evidence.

How to structure change-detection sampling

Define missions before the season starts. Think like a researcher or inventor, a scientist.

Identify the intervention zone. The matched control zone. A stability zone (control) which you do not expect to change. Visit all three with the same field protocol for data collection.

You will know whether the intervention worked, and whether anything else moved either side of the ideal which should not have.

Mix repeat sampling with repeat drone scans depending on the crop type, time of year, your goals and data metric type.

Quick Case Study: **Brendan's Drainage Payoff**

Brendan reworked a soggy corner of his Nectarine orchard by physical means - with ridges, and drainage installations - in an area that had nagged him for years.

One rainy season later, change detection sampling identified a lift in the canopy and NDRE values in the treated area. What a great feeling! Visual in-field inspections and sampling confirmed the results.

Precise drone data identified the area and confirmed his suspicions of chronic soggy ground; while adding quantitative insights in terms of the size of the affected area: the number of trees under stress, and how far into the block the drainage issue extended.

Repeat precision sampling over time (and strategic drone data monitoring) confirmed stronger trees, with improved stress levels, a better flush at springtime, excellent root growth - a marked improvement - and subterranean microbe activity not seen in years. Flowering was on par with the rest of the block earlier in the season, while leaf sampling showed favourable brix levels. With an improved fruit set and a problem corner taken care of, what more could he ask for?

The initial drone dataset provided hard proof where the issue was previously built on a hunch. This unlocked the budget for the drainage project, and the soil rejuvenation plan which brought about all these positive changes.

Drone info, plus change detection sampling turned gut feel into investment logic. More than that, he's got the greenlight to tackle the next set of orchards.

AI-Generated Waypoints: From Map to Route to Result

The sampling waypoint is where precision becomes practical. And nowadays, AI can actually help us to choose exactly the right trees and vines to visit.

Traditionally, AI is leveraged at the drone data processing stage for annotations and contouring, and to some extent with the on-platform filtering tools that help us to create zonal, linear, outlier, or cluster-based sampling routes.

Now, to take things a big step further, AI is used to quickly and easily generate the most representative trees to sample. No confusion, only a simple treasure map of waypoints without any guesswork.

Amazingly, this takes care of much of the analysis and planning work that is needed by us (the humans) for sampling route generation - all based on the drone data results and a blend of the metrics like canopy size and transpiration values.

The link to ROI is that this workflow not only guides teams to those trees with precision - but teams capture standardised, high-quality data every time by mobile app.

Soon, platforms will also optimise for minimal distance, logical entry and exit points, and safe paths that respect terrain and access.

Routes split by worker, set time boxes, and included special instructions for each waypoint (if needed) would become commonplace.

Offline sync capabilities keep routes and maps usable even without signal. RGB imagery overlays on apps keep routes clear and relevant (compared to older satellite visuals) so ground teams can find their way with ease.

Multispectral and thermal zone overlays add extra dimension to mobile app maps, while a dynamic waypoint interface provides sampling flexibility and clarity.

The *Aerobotics TrueFruit* mobile app is my preferred tool for this type of AI-generated sampling work, as it pairs seamlessly with the web-based platform which acts as the command deck for all precision sampling activities.

Standardised capture every time

Each sampling waypoint - or sampling mission at least - should carry a checklist protocol.

The required number of photos to be taken by the phone's camera. For yield sampling, is there a fruit diameter minimum - how soon can you start fruit size sampling after fruit-set?

Is there a fruit or nut count method - will you strip the sample trees or utilise a different non-destructive method?

Is there optional or mandatory leaf or soil sampling at these ai-generated trees?

Any barcodes or QR codes to scan for intake or sampling tracking? Traceability procedures or additional steps to take into account?

The more standard the protocol, the more comparable your datasets become across teams and seasons. So getting this "written in stone" is important, especially at the beginning of the season, and then to carry on with the plan throughout the season.

Quality control measures

With the mobile apps and platforms as mentioned earlier, supervisors can review samples in near real time - a massive leap in transparency for managers and stakeholders.

Missing photos, sample quality issues, or inconsistencies can be flagged and fixed the same day. This protects data integrity, while supporting continuity, which in turn protects and underpins data-backed decisions.

The power of feedback loops

The best part of AI-guided sampling is that it "learns".

Field outcomes and user feedback are able to be linked back to the trees and the base strata they came from.

The platform refines its sense of what is representative in your context. And over time, your sampling plans get sharper with just as little effort as always when it comes to route generation.

Quick Case Study:
Henry's Fruit Sampling Leverage

After a recent drone scan, Henry (a hands-on farm owner of a large citrus operation) decided to replace their manual fruit sizing and fruit count procedures with AI-backed waypoints and smart app data collection.

Time in the field fell from four months to pull in and consume orchard reports to just one 5-day work week for the entire operation.

Towards the end of the growing season, the harvest and packhouse managers scheduled staff and equipment movements with a level of confidence previously unknown.

The trucks ran full, not half-full. Teams picked on colour and size with less doubt and confusion. The marketing teams could inform on available size spreads in a timely fashion - without any thumbsucking - to both internal teams and external buyers and distribution partners.

Moreover, yield forecast error tightened from a wide band and inconsistency across cultivars and orchards, to a narrower, more favourable range of 3-5% error margins.

The change was not just an operational one. It was an astute financial decision. An AgTech investment that lowered risk, improved visibility, boosted confidence, smoothed communication, lowered risk, and built rapport with key stakeholders.

Without a doubt, Henry's one decision to leverage his drone data, AI, and clever mobile apps was a revolutionary step in Henry's agribusiness journey.

Timing and Phenology: When Sampling Speaks Loudest

The right method at the wrong time still disappoints. Tie sampling to stages where the signs you're hoping to find will likely be most visible.

Establishment and early growth: Pair zonal and outlier sampling to confirm survival, planting success, and early canopy development. This sets the baseline that future seasons will be measured against.

Fruit-set and early sizing: Cluster-based or zonal sampling gives a realistic early read on potential yield and fruit size distribution. It is the moment where logistics and input plans can still be shaped.

Mid-season stress windows: Combine thermal-informed outlier visits with NDRE checks to separate water-use issues from nutrient or disease constraints. This is where small adjustments still move the needle.

Pre-harvest validation: Walk representative missions to validate forecasts and check quality metrics. The goal is not only tonnes. It's pack-out, grade, and market timing.

Consistency is the multiplier: Revisit the same routes or waypoints at the same stages each year. You will build a time series that speaks clearly about progress and risk.

From Maps to Management: Turning Insight into Daily Practice

Precision sampling is not a simple move for report generation. It is a rhythm. Here is how the pieces fit together in a way that stands up during busy weeks.

Plan. Load your layers. Pick the method. Set standards. Generate waypoints. Check routes. Brief the team.

Walk the talk. Use the app the right way. Capture what is required. Add short notes where needed. Send feedback to your provider. And fix issues you can fix immediately.

Review results. Consume the data. Flag any anomalies. Decide what gets acted on now and what gets monitored as the season progresses.

Act on data. Adjust irrigation timing or pressure. Tweak nutrition in zones that need it. Set work orders for drainage or traffic management. Share before and after with your team so they see the point of the exercise.

Manage interventions. Organise harvest and packing teams. Brief management and marketing teams.

Record. Tag interventions to blocks and dates. The next survey will tell you if it worked.

The next set of sampling will update your database of insights, and your perspective on season status.

These steps, and the closed (and constant) feedback loops, is exactly how AgTech-based value compounds.

Putting It All Together

1. Fly a high-quality drone survey at a relevant phenological stage.

2. Create zones and clusters to sample.

3. Add outlier sets where the situation calls for it.

4. Generate AI routes with a realistic sampling target baked on capacity but with a focus on decent data collection volume.

5. Walk the plan. Capture data using standardised protocols.

6. Review data ASAP. Act on what you can fix quickly.

7. Record any interventions. Repeat at the next key stage. Stay focused and methodical.

You will feel the change within one season. Less wandering. Fewer arguments about where to go. More confidence that the numbers you present are solid. And a growing sense that your aerial layers and your field reality are finally talking to each other.

From Data to Decisions (Recap)

Precision sampling turns drone insights into operational clarity. Zonal methods capture the orchard's true range. Per-tree outlier checks diagnose what averages hide. Cluster-based sampling scales to large estates without losing signal. While historical change detection proves whether interventions have worked.

Finally, AI-generated waypoints tie it all to a sampling route that teams can walk straight away, with a level of representativeness and confidence seldom seen just a few years ago.

From sampling to strategy

Precision sampling turns drone layers into field decisions, and field decisions into measurable returns.

You now know that with a little tech and a measure of knowledge, you are able to choose (or be led with precision) on where to go, what to measure, and how to act on insights with confidence.

In the conclusion to follow, we pull a thread through the whole book, summarise the metrics that matter as a quick refresher, and focus on the biggest single habit that you should cultivate to convert drone data metrics and tools into real-world ROI.

PRECISION SAMPLING:
THE METHODS BEHIND MODERN ORCHARD SAMPLING

Conclusion:

Moving From Data to Decisions

Turning Insight into Outcome

This book has set out to help you turn aerial data into orchard outcomes. The goal was never technology for its own sake, but judgment sharpened by clearer information and repeatable, reliable workflows.

Metrics that Matter

RGB gives context and orientation, the honest picture of what is on the ground.

DEMs add shape and slope, the terrain signals that drive investigations and intervention around amendments, infrastructure changes, drainage efforts, erosion control, and access design.

Multispectral data, especially NDVI and NDRE, reveal physiological signals that our eyes cannot see, with NDRE adding sensitivity in dense canopies - invaluable for established plantings and younger blocks coming into production in their first few seasons.

Plant census statistics account for every asset in the ground, which underpins budgeting, insurance, replant planning, and precision sampling backed by artificial intelligence and advanced in-field mobile applications.

Canopy metrics quantify structure, growth, and variance - nutrition programmes, pruning regimes, and more - all which underpin orchard status improvements, variance management strategies, and eventually pack-out's that power farm-gate revenues.

Thermal metrics translate canopy temperature patterns into water use efficiency and transpiration uniformity insights, guiding irrigation fixes, and soil health programs before losses mount.

Each layer is useful on its own. Together, they form a diagnostic ecosystem the world of agri has never seen before. It's the drone-data vantage point. A lever to pull for those in the know.

From Scans to Systems

Maps are a starting point, not the finish line. The breakthrough is the workflow that follows the drone flight.

AI-guided sampling closes the loop between pixels in the air and action on the ground.

Zonal sampling captures the full range of performance in a block, while **per-tree outlier sampling** exposes what the averages hide. Together they give you both the shape of the orchard and the edges where problems or opportunities live.

Cluster-based sampling takes that logic to scale, letting large estates stay representative without losing signal. As well as offering a method of collection for younger blocks or smaller operations to ensure data volumes remain significant.

Historical change detection closes the loop by proving what worked and what didn't, so capital planning is driven by evidence, not hope.

GPS waypoints and app-guided sampling routes make in-field sampling or intervention plans walkable and repeatable, even when teams or seasons change, which means the workflow survives busy seasons and unpredictability while delivering consistent, comparable data.

ROI is the score that counts

Drone metrics pay for themselves when they cut waste, sharpen forecasts, protect quality, and de-risk large and important high-stakes decisions.

Managers can scale their inputs to canopy size or stress values variably, not to habit or tradition, or outdated advice. Overspending comes to a stop; and inputs are placed on hold where they've failed to equal desired results.

Sampling that truly represents the source orchards reinforces reporting. Forecasts tighten, and packhouses, marketing, logistics teams and leaders are able to plan with confidence.

Drone data and yield information combine with tailored action steps to lower variance and protect quality, because uniform orchards tend to deliver consistent size and grade - since orchard management on low variance blocks is that much easier.

On the investor front, one is able to de-risk capital by proving that interventions have moved the needle, then to use those results to justify the next project.

Treat the chase for ROI as a habit. Tag interventions, compare missions and action plans, and expect the numbers to show the value.

When Action Beats Intention

Farm managers that win do not wait for the perfect data. They understand that variance is part and parcel of farming. So they set standards, scan by drone at the right times, look for patterns, generate their in-field routes, and execute the sampling plan while recording results and fixes. Moving from data to decisions as swiftly as they can. There is little doubt that technology has made this last step much easier and streamlined than ever before.

Consistency and dedication turns insight into improvement. Improvement into progress.

But then again, if you have made it this far in the book, I am probably preaching to the converted.

Your Next 90 Days

1. **Pick one priority block** and run the full loop: Survey, build zones and outliers, generate AI routes, sample using a standard protocol, act, and review.

If you come unstuck on where and how to locate a drone service provider or AgTech platform that fits your needs - reach out - I am your new resource and it's one of my missions and privileges to be able to help you on your AgTech journey.

---

2. **Extend your efforts to two more blocks** with different challenges you are already aware of, but where perhaps you are not sure of the extent or severity.

3. **Create a simple quarterly scorecard** (3 months at a time) on canopy variance, NDRE distribution, a thermal uniformity or variance measures - including your action

plans - plus a short ROI note for each intervention. Remember phenology, your capacity and constraints, and keep things practical.

4. **Keep a running list of wins and still-to-solve items.** Log experiences and share feedback with your team and your providers, so progress stays visible and viable through conversation loops and continual improvements that move the needle over time.

Stay Current, Stay Sharp

The AgTech space moves quickly - particularly in the area of drone data. New sensors, better models, and improved workflows arrive every season.

For ongoing thoughts, ideas, case studies, and field notes, follow my blog at AgTech Diaries, and keep an eye out for future short-courses and publications that build on work in the field and with the farmers I support on a day-to-day basis.

The aim remains largely the same: a focus on practical and digital tools that help you turn data into decisions, and decisions into results.

In closing...

Remember, keep ROI front and centre by scaling inputs and efforts to orchard status and real-world needs, not according to things that "have always been done this way" or according to happenstance.

We live in an era of high-quality and ever-accessible and shareable data - with AI-backed systems and enhanced diagnostic techniques designed to make our farming lives more and more precise and efficient, powerful and impactful. It really is an amazing time to be alive.

So in closing, I want to leave you with a note on the biggest piece of value you can possibly leverage to propel you to the next level in your farming journey.

It's not the hardware, software, or any of those bells and whistles. Not the raw drone data, neither insights gleaned, nor outputs won or methodologies employed along the way.

It's the fervent and zealous disposition to take action on what you have experienced, what you have learnt, and what you know to be true about the next step in the journey to orchard optimisation.

The thinking starts here, but the real change starts when we *take action*.

Glossary of Terms

Clarity at a Glance

Whether you are reading the digital Kindle version of this book or one of the printed editions, speed matters.

Drone data has its own language, and some terms may feel unfamiliar at first. This glossary is a quick-reference guide in plain language. Use it, don't use it - it's here if you need it.

Think of the glossary as a pocket reference built into the book. Whether you are in the orchard with a phone or tablet, at your desk reviewing a scan report, a student, or simply connecting one concept to the next in your own time - this section might come in handy.

Each definition is short, clear, and practical. Use it as often as you need; that's exactly what it is here for.

Aerobotics
A precision agriculture company that uses aerial data and AI to provide insights for orchard crops, including health maps, fruit sizing, and yield risk reports.

Aerobotics Web Platform
The cloud-based software where drone data is processed, analysed, and displayed. Provides map layers, per-tree insights, yield estimates, and management tools accessible via browser.

Aerobotics Mobile Application
The companion app for field use. Syncs with the web platform to guide scouting, inspections, and sampling directly in the orchard.

AI Plant Detection
The use of artificial intelligence (AI) to automatically detect and contour individual plants in aerial imagery. Enables per-plant metrics like canopy size, health, and stress.

Canopy Area

The total surface area of tree canopies captured by aerial imaging. Used to monitor orchard growth, density, and tree health.

Canopy Volume

An indicative measure of the three-dimensional volume of a tree's canopy, derived from aerial imagery. Useful for assessing growth and productivity.

Computer Vision

A field of AI that trains computers to interpret and understand visual information. Powers automated plant detection, fruit sizing, and anomaly identification from drone imagery.

Compaction

The process by which soil becomes dense and hard, often due to machinery or foot traffic. Limits water infiltration and root development.

Data Layer

An individual dataset collected by drones or

sensors. Layers such as RGB, DEM, NDVI, or thermal can be stacked to provide richer analysis.

DEM (Digital Elevation Model)
A 3D map of the orchard's terrain generated from drone imagery. Used to analyse slope, drainage, and erosion risk.

Drainage
The natural or artificial removal of excess water from soil. DEM data often highlights poor drainage areas that impact orchard performance.

Drone Survey
A planned aerial mission where a drone captures imagery and sensor data across orchard blocks.

Evapotranspiration (ET)
The combined process of evaporation from the soil and transpiration from plants. ET data informs irrigation management.

Gap Analysis
Identifying missing or underperforming trees in an orchard through aerial census and canopy data.

Health / Stress Map
A visual layer (often NDVI or NDRE) highlighting zones of vigour versus stress in an orchard.

Hyperspectral Imaging
A remote sensing method capturing hundreds of spectral bands. More detailed than multispectral imaging, but costly and usually used in research.

Internet of Things (IoT)
A network of connected devices (such as soil probes, weather stations, and drones) that collect and share data for analysis.

Mobile Application (App)
Software designed for use on smartphones or tablets. In orchard farming, apps connect aerial data with in-field scouting and inspections.

Multispectral Imaging
Captures light across both visible and non-visible bands (e.g. near-infrared, red-edge). Generates indices like NDVI and NDRE that correlate with plant health.

NDRE (Normalised Difference Red Edge Index)
A vegetation index using near-infrared and red-edge wavelengths. Effective for detecting stress in dense, mature canopies.

NDVI (Normalised Difference Vegetation Index)
A vegetation index using near-infrared and red light. Commonly used to measure overall plant vigour, particularly in younger or sparse canopies.

Orthomosaic Map
A geo-referenced map created by stitching many overlapping drone images. Provides a high-resolution overview of orchards.

Orchard Variance
The degree of variability between trees in an orchard. Can be measured by canopy size, vigour, or yield indicators.

Phenology
The study of recurring plant life cycle events (e.g. flowering, fruiting, senescence). Important for timing surveys and management decisions.

Plant Census

A complete count of trees in an orchard, identifying gaps and mortality. Replaces manual counts with AI-based accuracy.

Pollination Mapping

Using aerial data to track bloom density and pollination success across orchards.

Prevailing Winds

Dominant wind direction in a region. Influences pollination, disease spread, and spray application.

RGB Imagery

Imagery captured using Red, Green, and Blue bands of light – the same visible spectrum our eyes see. Forms the foundation of drone mapping.

Scouting

Field inspections guided by aerial data. Focused on verifying anomalies or outliers flagged by drone metrics.

Smart Sampling

An AI-assisted approach to orchard sampling that

uses aerial data to identify representative trees, clusters, or zones for inspection.

Spray Drift
When applied chemicals move off-target due to wind or poor calibration. Precision mapping helps mitigate this risk.

Spray Drone
A drone designed to apply crop protection products or nutrients via aerial spraying. Distinct from survey drones.

Thermal Imagery
Imagery capturing heat signatures. Used to assess plant transpiration, irrigation performance, and stress.

Transpiration
The release of water vapour from plants. Plays a central role in cooling leaves and regulating plant function.

Transpiration Uniformity (TU)
A thermal metric that shows how evenly trees across an orchard are transpiring.

Transpiration Variance (TV)
Measures differences in transpiration between trees. High variance indicates uneven irrigation or stress.

UAV / UGV
Unmanned Aerial Vehicle (drone) or Unmanned Ground Vehicle. Used in precision agriculture for data collection or interventions.

User Interface (UI)
The design of software dashboards or apps that enable users to interact with drone data outputs.

Variable Rate Application (VRA)
Applying inputs such as fertiliser, water, or sprays in variable amounts based on mapped data, improving efficiency and reducing waste.

VTOL (Vertical Take-Off and Landing Drone)
A drone that combines multi-rotor take-off with

fixed-wing flight capability. Suitable for larger orchards with varied terrain.

Wireless Technologies
Communication methods such as Wi-Fi, LoRa, or cellular networks that connect field devices, sensors, and cloud platforms.

Yield Risk Report (YRR)
A data-driven report from *Aerobotics* combining aerial and AI analysis to estimate orchard yield risks.

Zonal Layer
A data output that groups trees or orchard blocks into zones based on similar characteristics (e.g. vigour, canopy size, or stress).

Acknowledgements

First off, thank you to my family and for your patience while I utilised many of our early morning and late night family time hours to make this book happen.

To my son Joshua, thank you for all the shared time and for accompanying me at my desk during the final few chapters - always reminding me of why I started this project in the first place.

Secondly, none of this would have actually been possible without the opportunity to join the force that is *Aerobotics*. Thank you to founders James and Benji for the opportunity to serve on the team, and for your support.

On that note, thank you to the *Aerobotics* colleagues who have supported my enthusiasm for the Drone Scan product suite - most notably Liaan Janse van Vuuren, Saadiq Jacobs, and Brett Agiotis.

Although many colleagues not mentioned here have impacted my journey in a positive way, whether they realise it or not, I want to extend a general gesture of gratitude - thank you to everyone.

Aerobotics has been on the whole a great place to be curious about tech and processes in the SaaS agri space. More acutely, exploring what is possible with drone data has been exciting - and helping farmers with digital tools and solutions has been gratifying to say the least. Thank you again to James, Benji, and my manager and workplace confidant JC vd Walt for your support while on this journey.

Of course, I owe a debt of gratitude to everyone in the agri ecosystem I have worked with. From farmers and agronomists, to pilots and managers, thanks for welcoming me into your offices and orchards. We've shared *bakkie* rides, *braai* time, and many chats about life, often over a hearty meal, or a good cup of coffee. It's been great.

Also, thank you to all the farmers and managers for being so gracious and welcoming, even when I showed up with my family (or my dog as my sidekick) during my fair share of travels to both nearby and far-flung operations. Your hospitality, honourable characters, and openness to share, together with your honesty about real world challenges has shaped me - and much of what you'll find in my writing.

Please know that my hope is that this little book will become one of many. Catalysts that steadily work to shift farmers toward a smarter way of mapping, monitoring, and mobilising food production efforts.

Also, a thank-you needs to go out to friends, Mark Blew, Chris Cherry, Derrik du Toit, Graham Fawkes, Paddy Gloster, Gert Laubscher, Marco Matteucci, Craig Miller, Chiliboy Ralepelle, Chris Hooton, and Angelo Nel for your quiet yet steady support.

And lastly, I would also like to present a special nod of gratitude to those who have questioned my madness.

In general, I've learnt that when everyone agrees with you, it's usually a bad sign. So thank you for keeping me on my toes and for confirming that I should probably keep going.

All the best,
Ken

About the Author

Ken Treloar is the director of *AgTech Publications (Pty) Ltd* and the voice behind the The AgTech Diaries blog. His work sits at the intersection of technology and farming, producing practical resources that help producers and industry professionals turn aerial data into informed action.

Drawing on his horticultural background and business management credentials, plus years of experience with drone technologies and precision agriculture, Ken writes with a clear purpose: to make complex tools accessible and valuable, rooted in the realities of perennial crop farming.

Ken's books and courses serve as guides for agricultural decision-makers seeking both immediate improvements and long-term return on investments.

When not writing, presenting, consulting, or teaching; you can find Ken at *Aerobotics* as part of the customer success team, supporting farmers all over the world.

Discover more of Ken's written works and resources at *linktr.ee/agtechdiaries.*

For consultations, speaking opportunities, and strategic collaborations, you can reach Ken via email at: *agtech@kentreloar.com*

Continue the Journey

Drone Data Metrics for Orchard Farming is part of a broader effort to make agricultural technologies way more practical, accessible, and impactful.

- Explore additional posts, books, courses, and guides at: *www.agtechdiaries.com*
- Connect with Ken on LinkedIn: *linkedin.com/in/kentreloar*

Remember, the journey toward smarter and sustainable farming continues... and every decision counts. The right type of insights, and the will to step up to the challenge, makes all the difference.

THE AGTECH DIARIES
AGTECH - DRONETECH - AGRIBUSINESS

AgTech Publications (Pty) Ltd

What are you waiting for?
Get started.

AGTECH
PUBLICATIONS

EDUCATE ✦ CULTIVATE ✦ ELEVATE

www.ingramcontent.com/pod-product-compliance
Lightning Source LLC
LaVergne TN
LVHW052250200926
848481LV00002B/326
* 9 7 8 1 0 4 9 2 2 7 6 1 0 *